George F. Wright

Greenland Icefields and Life in the North Atlantic

With a new discussion of the causes of the ice age. Vol. 1

George F. Wright

Greenland Icefields and Life in the North Atlantic
With a new discussion of the causes of the ice age. Vol. 1

ISBN/EAN: 9783337316969

Printed in Europe, USA, Canada, Australia, Japan

Cover: Foto ©berggeist007 / pixelio.de

More available books at **www.hansebooks.com**

GREENLAND ICEFIELDS

AND

LIFE IN THE NORTH ATLANTIC

WITH A NEW DISCUSSION OF
THE CAUSES OF THE ICE AGE

BY

G. FREDERICK WRIGHT, D.D., LL.D., F.G.S.A.

AUTHOR OF THE ICE AGE IN NORTH AMERICA, ETC.

AND

WARREN UPHAM, A.M., F.G.S.A.

LATE OF THE GEOLOGICAL SURVEY OF NEW HAMPSHIRE, MINNESOTA,
AND THE UNITED STATES

WITH NUMEROUS MAPS AND ILLUSTRATIONS

NEW YORK
D. APPLETON AND COMPANY
1896

LIST OF ILLUSTRATIONS.

xiii

PREFACE.

THE immediate impulse to the preparation of this volume arose in connection with a trip to Greenland taken on the steamer Miranda with an excursion party organized by Dr. F. A. Cook in the summer of 1894. While preparing to make the most of this excursion, much difficulty was encountered in collecting the facts which one would most like to know concerning this mysterious land. The varied and exciting fortunes of the Miranda, while not sufficient to form the framework of a volume, were still of no small value in giving vividness to one's conceptions of the unique conditions of the country, enabling one who shared them to enter with better understanding into the descriptions given by others, and to combine them into a more satisfactory general view. Upon some points, also, our observations furnished a positive enlargement to our knowledge of the country. Since, therefore, numerous references will be made to incidents in our voyage, it will be profitable here to give a brief sketch of the fortunes of the expedition.

On the 7th of July, 1894, the party, consisting of fifty-one, besides the officers and crew, set sail from New York in the Miranda, an iron steamship two hundred and twenty feet long and of eleven hundred tons burden. After stopping at Sydney, Nova Scotia, and St. John's, Newfoundland, we steamed out of the latter place on the evening of July 15th, aiming to touch upon the coast of Labrador, to leave a portion of our party, and a few native Eskimos who had been at the Columbian Exposition.

Everything went well up to the morning of the 17th, when, during a dense fog, the steamer ran directly into an iceberg; but the injury proving to be entirely above the water line, we made our way to Cape Charles Harbour, on the coast of Labrador, about fifteen miles distant, where our engineer and carpenter were able to make such temporary repairs that we could return to St. John's and readjust our plans. Here new iron plates were substituted for those which had been injured, and everything was put in good condition, so that on the 28th of July we decided to sail directly for southern Greenland. On the 3d of August the mountains in the vicinity of Frederikshaab came in view, but ice prevented our reaching land. Steaming slowly west and north as the fog permitted us until the morning of August 7th, we found a clear passage open to a broad bay, which proved to be the harbour of Sukkertoppen, the largest, and in many respects the most interesting, of the Eskimo settlements upon the coast.

The captain decided to remain here two days, which afforded me, with a small party, opportunity to make an excursion in small boats a considerable distance up one of the fiords which set back from this point toward the inland ice. On the morning of the 9th we started for the north, hoping to reach Disco and to visit the Jakobshavn glacier; but it was now too late in the season to think of carrying out our original plan of going still farther north, and reaching Peary's headquarters in Inglefield Gulf. And, now, even these more modest hopes were rudely dashed, by our steamer's running upon a reef about seven miles out, compelling us to return to Sukkertoppen in a disabled condition. In certain conditions of the water we should have passed safely over, but heavy swells caused the steamer to strike the rocks several times, and injured the bottom under the ballast tank near the middle of the vessel.

Hearing that some fishing schooners had been seen somewhere in the vicinity of Holsteinborg, small boats were despatched in that direction for assistance. As ten days must elapse before their return would be possible, opportunity was given me for a longer and more satisfactory expedition into the interior. The time was improved by an excursion to Ikamiut, a miserable Eskimo settlement a few miles from the head of a fiord into which there enters a great glacier projecting from the inland ice.

Upon returning to Sukkertoppen, the schooner Rigel, under command of Captain George W. Dixon, of Glou-

cester, Massachusetts, came to our relief. As the exact
amount of damage to the Miranda was difficult to determine, the captain entertained the hope that she might
be able to steam back to Labrador, but the hazard was
too great to venture without a convoy. The members
of the party were therefore all transferred to the Rigel,
which was only a hundred feet long and of a hundred
tons burden. Everything but the bare necessities was
left upon the Miranda. We were, of course, very much
crowded upon the schooner, but the captain and crew
were so cordial in their services, and the relief was so
grateful, that every one was contented and happy. The
prospect of having to winter in Greenland was the less
endurable from the fact that our provisions would be
sufficient for only about two months; and the stock
of civilized food in the place was so small, that before
the middle of winter we should all be reduced to the
native diet of uncooked fish, raw meat, and blubber.

On the 21st of August we set out for home, with the
Miranda taking the Rigel in tow, but at the end of
thirty-six hours, at midnight, when we were 300 miles
from land and the sea was rolling heavily, the leak in
the Miranda started anew, and she sounded the signal
of distress, informing us that it was necessary to abandon her, and that we should lie to with the cable with
which we were towed still attached until the officers
and the crew could be transferred to the schooner.

The scenes of the next few hours were the experience
of a lifetime. With a man standing ready to cut the

cable in case the Miranda should suddenly go down, we tossed about upon the waves for three hours, awaiting the approach of the first boat load from the steamer. At length it arrived, and, after the exercise of an infinite amount of labor and skill, was safely brought alongside, and the men upon it transferred to the deck of our stanch schooner. Others followed, until after another hour and a half we heaved a sigh of relief when the last man was on board and not a life had been lost. On leaving the Miranda, Captain Farrell had shipped the cable, so that she could drift away from us, and we took our last look at the unfortunate vessel. The lights were still burning, the smoke was issuing from her chimneys, the pumps were still vigorously at work, and the wheel was still turning. Thus, with nearly all our personal effects and treasures on board, she steamed off to her lonely fate in the mist and darkness of that boisterous sea, while we directed our course for the southern coast of Labrador.

There were now ninety-one of us on board the little schooner, which was calculated for the accommodation of only twenty. After twelve days of various vicissitudes we reached Sydney in safety, and the perils and discomforts of our journey were over. So far as our experiences enabled us to add to the knowledge of the regions visited, and seemed to be of general interest, they have in their appropriate place been worked into the various chapters of the book.

In the preparation of the volume I have been happy

in having the aid of Mr. Warren Upham, with whom
I have been closely associated in glacial investigations
ever since we worked together on the New Hampshire
Geological Survey in 1876. His continuous labours in
investigating the glacial phenomena of New England
and the Northwest, his minute knowledge of natural
history, and his wide study of the literature of the subject, give great weight to the conclusions to which he
has arrived upon the theoretical questions connected with
the Glacial period and its relation to existing ice-fields.
Mr. Upham has drafted the series of maps here used
and has prepared chapters viii to xiv, but in every part
of the volume we have co-operated with each other, and
availed ourselves of each other's knowledge.

In conclusion, I must express my obligations to Professor William H. Brewer, and Messrs. George W. Gardner, Walter S. Root, A. R. Thompson, James D. Dewell,
R. Kersting, J. R. Fordyce, and Drs. R. O. Stebbins,
F. A. Cook, and J. F. Vallé, for the use of photographs
taken by them during our excursion. The illustrations
obtained from Professor T. C. Chamberlin were first
published in the Bulletin of the Geological Society of
America for February, 1895, and are used by permission of that society.

Grateful acknowledgements are due, also, to Judge
Henry C. White, of Cleveland, Ohio, for the use of
his large collection of books upon Arctic and Antarctic
explorations.   G. FREDERICK WRIGHT.

OBERLIN, OHIO, *November, 1895.*

CONTENTS.

2

Fig. 1.—Iceberg stranded at the entrance of St. John's Harbour, July 15, 1894.

Newfoundland stood out sharply upon the horizon from morning till night; but most inspiring of all was the constant procession of icebergs which the steamship Miranda and her passengers were meeting the entire day. The size of many of these bergs was enormous, and their shapes were often fantastic and beautiful in the extreme. One, which we attempted to measure, was estimated by the best judges to have pinnacles which rose more than seven hundred feet above the water. The area of its base must have been as much as ten or twelve acres, or as large as that of the largest of the pyramids of Egypt. For more than thirty miles this huge object continued to tower upon our vision in lonely solitude far up above the watery horizon. If the shape of it had been regular, this would have implied an enormous depth below the water, since the specific gravity of glacial ice is such that about seven cubic feet are below water to one above; but in this case the visible part of the berg was much wasted by the joint action of rain, sun, wind, and waves, while the submerged base was greatly extended as compared with the portion which was above the water. What at first seemed to be one gigantic pyramid, proved, as we shifted our position, to be two or three towers, separated by long spaces, yet rising from a common broad base of blue ice below the water. Over this submerged portion the waves were dashing as on a sunken reef of rocks. The general appearance reminded one of the ruined cathedral at Utrecht, whose tower stands on one side of the street and the choir upon the other, the vast nave having disappeared, as the result of some accident, centuries ago. In the course of the day bergs of every imaginable form passed by us, or rather we passed by them, for their motion was inconsiderable. Their beauty of color was also indescribably diversified. The surface was the pure white of newly fallen snow; the perpendicular face, where fresh, was the deepest azure, and where not fresh

FIG. 2.—The most majestic iceberg seen off the coast of Labrador.

was intersected by numerous seams of blue; while the base, where partly obscured by water, shaded into a delicate green.

Sixty bergs of large size were sometimes in view from the deck of our steamer at once. They were specially numerous near the Strait of Belle Isle, toward which the Labrador current is attracted by the opening into the Gulf of St. Lawrence. In this vicinity they are always a serious obstacle to navigation; while the danger is increased during the early part of the summer by the presence of extensive masses of floe ice, which, though less conspicuous, are more dangerous than their gigantic companions. On the island of Belle Isle the Canadian Government has built a lighthouse—the most northern on the eastern coast of America—and has established life-saving stations and stored supplies of food,

FIG. 3.—A more distant view of the same.

both to encourage commerce to take the most direct route from the St. Lawrence to Liverpool, and to afford relief to the many hapless navigators that are sure to meet with disaster in those treacherous waters.

A friend who was sailing from Quebec to Liverpool in the early part of July, 1884, informs me that his steamer was detained in the ice for two or three days near the entrance to the strait; while another, who had taken passage from Glasgow to St. John's, Newfoundland, on one of the Allan line of steamers, was caught amid the icebergs in a fog about one hundred miles east of the latter place, and detained in that hazardous condition from Monday until Thursday before the weather cleared sufficiently to render it safe to complete the passage. These experiences were by no means extraordinary.

On the morning of the 17th we were destined to witness a different phase of arctic navigation from that previously enjoyed. We were in the mouth of the Strait of Belle Isle, and had just passed its lighthouse. Icebergs as vast and beautiful and numerous as ever were around us; but their beauty, and in most cases their presence, was obscured by a fitful but dense fog. Still the chances of encountering one seemed so slight, that we continued on our course, though at slackened speed. Suddenly a large spot in the mist ahead of the ship lighted up as if we were coming to a clear space. But to the experienced eyes of the mate and of the ice-pilot it was an ominous spectre, for it was the "ice blink" of a huge berg, which almost at that very instant emerged from the mist, towering hundreds of feet above us, and stretching out many hundred feet on either side. It being too late to avoid it, safety, if it could be secured at all, lay in taking the collision straight ahead, and the rudder was turned accordingly at the same instant that the wheel was reversed. The collision came all too soon. Great masses of ice fell upon the deck. The steamer reeled like a drunken man. But the passengers scarcely had time to secure an upright position again before all was quiet as death, while we anxiously watched the carpenter as he sounded the wells to see if there was a leak. Meantime the huge berg gleamed down upon us from its serene height in the mist, and revealed clearly the

painted sides of the great gash which had been made in
it by the bow of our iron ship.

After a few minutes of breathless anxiety it was
ascertained that, beyond the breaking of three or four
iron plates of the ship high above the water line and
the disabling of one anchor, there was no damage done.
Fortunately, we had hit this particular iceberg where
there was no projecting foot below the water; otherwise it would most surely have been a fatal collision to
us.   It was certainly thrilling for us to reflect that we
had sailed, or steamed, by a most circuitous route for
more than one thousand miles to encounter in the mist
this particular berg, and had hit it at almost the only
safe point which it presented for attack.

After the collision, the comparative safety of iron
and wooden vessels in encountering icebergs was a subject of much animated discussion on board.   Our own
actual immunity from absolute disaster was an argument
in favor of iron, while the pilot assured us that a wooden
vessel in such a collision would have had her bowsprit
driven in so as completely to disable the ship and send
her to the bottom.   The worst disaster of the season
along this coast had occurred a few weeks before to a
wooden ship which encountered the ice not far from the
locality of our own accident, but with the most serious
consequences.   The ship sank almost immediately, and
the crew and passengers, consisting of men, women, and
children on their way from Newfoundland to the summer fishing stations in Labrador, were many of them
drowned, while the others were rescued with great difficulty from cakes of ice on which they had taken refuge.

One pretty constant feature in the appearance of the
largest icebergs could not fail to attract our attention.
The most of them seemed to be partly turned over, so

that the strata of ice which were originally horizontal were dipping at a considerable angle with the horizon. The natural explanation seems to be, that in the long journey which these icebergs have taken they have presented, through almost the entire course, the same side to the sun, so that the southern exposure has constantly melted more rapidly than the northern side, thus disturbing the equilibrium and allowing the centre of gravity to migrate toward the northern edge, which con-

Fig. 4.—An iceberg which has shifted its plane of equilibrium.

stantly settles deeper into the water at the same time that the southern is elevated. Eventually the berg turns upon end; thus producing a sudden transformation in appearance.

The history of these objects, the distance through which they have drifted, and the rate of their motion, are questions of much speculative interest. From all accounts it appears that the large bergs are formed by the breaking off of masses of ice from glaciers whose ends project into the deep water of the sea, and so are composed of innumerable strata of consolidated snow which has accumulated upon some continental area. So far as we know, also, few icebergs of large size are derived from glaciers upon the American side of Davis Strait. Those of the Labrador current represent for the most part, therefore, the wastage from the Greenland glaciers, and are thus impressive witnesses to the extent of the glacial accumulations upon that desolate land.

For the most part, also, the icebergs of the North Atlantic enter Baffin Bay north of Disco Island. A due quota comes from the glaciers at the head of Umanak Fiord and from the comparatively unknown region about Melville Bay, while others move in from Smith Sound, and from points still farther north. The journey, therefore, which most of those seen off the coast of southern Labrador have made covers a distance of eighteen hundred miles, while those which reach the latitude of Washington are no less than twenty-five hundred miles from their starting-place. The length of this journey is a witness at once to the vast mass of ice in the bergs and to the initial low temperature of the water which floats them southward. The ice, however, does its part in perpetuating the coolness of the Labrador current, the modifying influence of whose temperature is felt upon the coast as far south as Cape Cod.

The continuance of such a procession of icebergs as floats southward with the Labrador current is due to a combination of causes. The snowfall over the vast continent of Greenland is not excessive, representing probably no more than from ten to fifteen inches annually when reduced to the standard of water, but even this

small amount exceeds the melting capacity of the sun's
rays in that latitude; so that there is an annual addition
to the thickness of the ice-sheet. This excess of frozen
water is enabled, through the plastic character of ice, to
work off slowly toward the sea level in lines of least re-
sistance, which are determined largely by the character
of the Greenland coast. Nearly everywhere this is
mountainous, with occasional openings to the sea con-
stituting the natural lines of drainage which have doubtless been occupied by water courses from time imme-
morial. In general, these openings are narrow and deep,
corresponding to the fiords of Norway. The movement
of the inland ice is toward the head of these fiords,
where the frozen current becomes concentrated, and
pushes out into the deep water, presenting everywhere a
perpendicular face of from one hundred to two or three
hundred feet in height, and extending from cliff to cliff
across the fiord.

The masses of ice from the front of these glaciers
are broken off in two ways. Much falls off directly into
the water from the upper edge of the precipitous face;
for the movement of the ice at the top is faster than
that at the bottom, and so tends constantly to project it
forward until it overhangs the lower portion and falls
by the mere weight of gravity. But such fragments are
comparatively small. The larger bergs are formed when
the glacier has been pushed so far out into the water as to
be lifted by its floating power, and so separated from the
parent mass. It is thought by Rink that the end of a
glacier may sometimes be pushed forward for several
miles after its bottom has become separated from the
bottom of the fiord, and before the buoyant power of
the water is sufficient to cast it off as an iceberg.

One of the glaciers coming into Disco Bay, observed

by Helland, was estimated to have a thickness of 920 feet and a breadth of 18,400 feet, and was found to be moving forty-seven feet per day. At this rate, 300,-000,000,000 cubic feet of ice would be sent off by this glacier in a single year. Another glacier surveyed by

FIG. 5.—A symmetrical iceberg, probably two hundred and fifty feet high.

the same authority sends off 79,000,000,000 cubic feet annually. Rink estimates that on the Danish part of the west coast of Greenland up to 74° N. latitude there are twenty ice-fiords from which bergs issue, and that 120,000 square miles of territory contribute to furnish the supply. To provide the amount of ice carried away by these floating bergs, an excess of only two inches and a half of snow over the contributing area would be requisite, so vast is the cumulative effect of apparently insignificant causes when sufficient time is at command.

As already remarked, the Labrador current bears also an immense amount of floe ice. This, too, has been formed along the shores of Baffin Bay, and is from

time to time separated from them to accompany the
procession of icebergs southward. Floe ice rises but a
few feet above the surface of the water, except in some
places where the cakes have been piled one upon another in the collisions which occasionally occur during
storms.

The westerly movement of this ice deserves attention.
According to a well-known law, currents which flow
from the north to the south are thrown to the west by
the revolution of the earth upon its axis, since the easterly motion of the earth's surface increases with each
parellel of lower latitude. As a consequence, this floe
ice, together with many icebergs, is so crowded against
the coast of Labrador as greatly to interfere with navigation to its ports. Oftentimes a whole summer passes
during which it is almost impossible to enter any of the
northern ports on account of this ice, and sometimes it
is difficult to get into any of the ports even as far south
as Hamilton Inlet until past the middle of summer.

In the main, the course of the icebergs and the floe
ice is the same. But the great depth to which the icebergs sink places them under the control of the under
currents, and makes them independent of wind and tide.
It is not unusual, therefore, especially in the far north,
to see a towering berg moving directly against the wind
and tide, and crashing through the thinner masses of
floe ice which encumber the surface of the water. Dr.
Kane records that at times he availed himself of this
mode of locomotion, anchoring his ship to an iceberg
which was moving northward when all the surface elements conspired to drive him southward.

But to make up for the inconveniences which the
ice of the Labrador current occasions to the meagre
commerce of the region, it brings along almost the only

booty which secures the country any commerce at all.
In the early spring the saddleback seals (*Phoca Grœnlandica*) * of the far north move southward in vast
numbers with this ice to propagate their young in the
latitude of southern Labrador. Naturally, also, the
polar bear avails himself of the same means of locomotion to keep company with the seals, which constitute
his favourite food. During April and the early part of
May numerous steamers, fitted for the purpose, set out
from St. John's, Newfoundland, and venture boldly into
this ice of the Labrador coast to secure the game thus
brought within their reach.

One of the first objects which attracted our attention
upon entering Cape Charles Harbour for repairs was the
magnificent skin of a polar bear which had been killed
the previous season on the land near by. Thinking he
had moved far enough south, the brute had deserted the
floe, and, having reached the land, immediately turned
his steps northward. The spring was already well
advanced and the snow was soft, so that at every step
Bruin went in the full length of his legs. It was an
easy matter, therefore, for the hunters, inexperienced
though they were in dealing with such large game, to
follow him upon snowshoes and secure his capture.

Three remarkable experiences in the floe ice of
Baffin Bay give us much definite information concerning the motion of the current which bears it southward.

On the 8th of May, 1854, Sir Edward Belcher, when
in search of Sir John Franklin, abandoned one of his
ships, the Resolute, on Beechy Island, in Barrow's
Straits, about latitude 75° N. and longitude 95° W. from

---

* This is not to be confounded with the fur seal (*Callorhinus
ursinus*) of Behring Sea.

Greenwich. Upon leaving the ship everything was put in order, the rudder was taken on board, and "every movable packed away below or securely lashed on deck." A year and four months later, on the 10th of September, 1855, the Resolute was sighted by Captain Buddington in the ice pack off Cape Mercy, near the mouth of Cumberland Sound, in latitude 67° N., having drifted through Lancaster Sound and down Baffin Bay, a distance of eleven hundred miles, in the sixteen months which had elapsed since its abandonment. The ship was still in seaworthy condition, so that a crew was put upon it which brought it safely to New London, Conn. Appropriately, Congress rewarded the captain and his crew with a grant of forty thousand dollars, and had the ship repaired and sent to the English Government as a token of national good feeling.

The second experience in the Labrador ice current upon which we pause is a chapter from McClintock's narrative of the discovery of the fate of Sir John Franklin. McClintock and his party had sailed on July 1, 1857, from Aberdeen, Scotland, in the Fox, a screw yacht of one hundred and seventy-seven tons burden. On the 17th of August the vessel became encumbered in the ice pack of Melville Bay, and, being unable to extricate themselves, the party was obliged to await the rigours of an arctic winter amid such unpropitious surroundings. At first the drift was westward, and they were so near the land that they could see it all around Melville Bay from Cape Walker to Cape York. In fourteen days they had drifted forty miles. On the 16th of September they were within twenty-five miles of Cape York, and within twelve or fifteen miles of open water in that direction; but it was too late to move. New ice was forming about them, and they were com-

pelled to prepare to winter in the ice pack. On the
26th of September, " Snowy Peak, to the north of Melville Bay and ninety miles distant, was still in view."
The winter passed with little to break the dull monotony except an occasional storm or the appearance, now
and then, of a polar bear on the floe. During December
they drifted sixty-seven miles directly down Baffin Bay
and were in latitude 74°. On January 17th they were
within one hundred and fifteen miles of Upernivik,
having drifted sixty miles during the first half of the
month.

On the 7th of March they were so far south and east
that the highlands of Disco were visible ninety miles
away. On the 26th and 27th of March they drifted
thirty-nine miles. On the 26th of April the ice of
the pack had become so much loosened that the Fox was
able to free herself from her long imprisonment. The log
reads : " During our two hundred and forty-two days in
the pack ice of Baffin Bay and Davis Strait we were
drifted 1,194 miles geographical, or 1,385 statute miles.
It is the longest drift I know of, and our winter, as a
whole, may be considered as having been mild but very
windy " (p. 99).

Instead of going home after this experience, McClintock and his crew sailed to Holsteinborg, on the coast of
Greenland, and, having repaired their vessel, set out
anew upon their errand and were successful both in
finding the last relics of Sir John Franklin and in definitely determining the fate of his expedition. The Fox
is still (1894) doing good service in Greenland waters,
being employed at Ivigtut in towing barges for the Cryolite Mining Company.

The third adventure surpasses all others in dramatic
interest. Captain Hall, with a well-equipped party upon

3

the Polaris, had sailed from New London, Conn., July 3, 1871, upon an arctic exploring expedition. On August 30th he reached 82° 29', his most northern point, where he began to drift south. Steaming out of the pack and finding a harbour on September 3d, Hall soon after made an extensive sledge journey, from which he returned only to die on the 21st of October, when the command fell upon Captain Buddington, whose name has already been mentioned in connection with the sighting of the Resolute. The winter and the following summer were spent by the crew in these high latitudes in prosecuting the objects of the expedition; but about the middle of September, 1872, the ship became fastened in the ice pack, in latitude 79° 34'. On the 15th of October the ship was so badly damaged and in such imminent peril that provisions and stores were thrown out upon the ice, where a portion of the party also took refuge. At half past nine in the evening the ice cracked so as to liberate the ship and loosen its ice-anchors, and in the darkness of a stormy night the two parties were separated.

On the floe were nineteen persons—"Captain Tyson, Mr. Meyer, the meteorologist, the steward, the cook, six seamen, and the Eskimos Joe and Hans, with their wives and children, including a baby born to Hans August 12th, and then christened Charles Polaris." Several of the floe party passed the dark hours of the first night on separate pieces of ice which had been broken off from the main floe, but they were all brought together upon the following day. Tyson naturally took command of the party, and proved himself adequate to the situation. The separation occurred in latitude 78° N., in the entrance to Smith Sound, not far from Littleton Island.

The experiences of the party are told in various diaries kept by different members, and in answer to questions of the Congressional Committee which investigated the conduct of the expedition, all of which will be found in full in the reports of the Secretary of the Navy and Postmaster-General, First Session, Forty-third Congress, 1873–'74. It is necessary to give only the briefest summary of the events, but even that is stranger and more interesting than fiction.

After gathering together upon one floe, the party took stock of their possessions and found themselves with " two boats, some clothes-bags and muskox skins, fourteen cans of pemmican, fourteen hams, some canned meat, a small bag of chocolate, the tent built on the floe previously, and twelve bags of hard bread therein ; besides an 'A' tent, instruments, chronometer, etc." The ice was so broken and unsteady that a continual removal of stores was necessary. All worked hard until about twelve o'clock at night, and then, exhausted, lay down to rest amid drifting snow and a fearful tempest. All the papers and records were lost.

Next morning, October 16th, they found themselves wedged in between an iceberg and land which they could not reach. During the day the ship was seen under full steam, but they were unable to communicate with her, and were compelled to resign themselves to their fate upon the ice. Through the following day the gale continued, and the ice kept breaking off upon the edges, so that only a small piece was left them. The provisions were estimated to be sufficient to last four months, at the rate of three quarters of a pound per day to a man. They had no fuel for fire, except as the two Eskimos might be fortunate enough to kill seals to furnish them with blubber for a lamp. On October 22d three snow huts were built, and three seals had been shot, enabling them to make soup over a lamp. On the 23d, a missing boat, and a plank house from which they had been separated by the storm, were discovered. The next few days were occupied in securing the boat and the house and the provisions that were in it. Among the additions to their party at this time were two dogs that had been separated from them on other floes.

On November 3d they were found to be actually adrift, so that all hope of getting back to the Polaris was given up, and they settled down to await their fate. On the 6th they saw still more clearly that they were drifting south and west. On that day the Eskimos caught a seal, upon whose uncooked meat they made a full meal. On the 10th they were drifting rapidly, and passed the Carey Islands, but the Eskimos had had no success in hunting. On November 21st two seals were caught, and permanent snow huts were erected. November 22d was calm and clear, and brought them one seal. The floe then was surrounded with more or less water. On the 23d another seal was shot. Thanksgiving Day, November 28th, witnessed no extension to their larder, but was celebrated by a slight change in diet.

December passed with very little to break the monotony, except the shifting of the thermometer up and down the scale between $-26°$ and 0, accompanied with changes in the wind and alternations in the cloudiness of the skies and in the brilliancy of the auroral displays. On December 21st it was light enough to read print at ten o'clock in the morning. On December 25th they had a Christmas dinner of two biscuits, half a pound of ham, and a cup of blood soup each. Little game was shot during this month.

January passed also with little variation in experiences, except that the thermometer ranged from 10° to 15° lower than in the previous months, reaching by the 13th —40°. Gales and snowdrifts were also increasing incidents of their life, while the Eskimos shot only an occasional seal, which barely helped them to eke out their scanty supply of fuel and keep from freezing. On the 20th they were in latitude 70° N.

February passed with slightly higher temperature but an increased amount of stormy and cloudy weather. On February 26th the allowance of food was reduced to seven ounces per day.

March was inaugurated by a temperature of —34° and the shooting of sixty-five dovekies. These little birds formed an important part of their additions to the larder during the month. From March 9th to the 12th the floe cracked badly, and they held themselves in readiness to take to their boats in case of disaster; but though the floe was completely broken up, the piece of ice they were on was left intact. On the 27th a bear was killed, and on the 31st four seals. They had then reached latitude 59° 41'.

On April 1st they left their snow encampment and proceeded to the southwest in their boat. Seals were plenty, and they were well supplied with provisions, but the inconvenience of hauling the boat upon the ice floe was great, and hazards of every sort increased. On April 5th a great gale set in from the northeast, breaking off pieces of the ice upon which they had taken refuge. This gale continued until the 9th, when a heavy sea was breaking over them, and they were compelled to stand by the boat until twelve o'clock to keep it from washing off, the children being in the boat. With varying experiences between life and death, they pressed onward until the 19th of April, when they were obliged to remain by their boat on a piece of ice during the entire night, while the sea washed over them, and fragments of ice were pelting at their feet.

Thus they kept on until the 29th of April, when at daylight they sighted a steamer five miles off. Launching their boat amid the floating ice, they made for her, but soon got fast and could go no farther. Landing upon a piece of ice, they hoisted their colours from an elevated place, fired their rifles and pistols, and heard what they supposed were return shots from the steamer. But in the afternoon the steamer turned away from them. It seems that the signals of the wrecked party had not reached the steamer. The firing which they had heard was from the seal hunters out upon the ice. But at five o'clock on the morning of April 30th, when the fog cleared away, it disclosed another steamer near by them. This proved to be the Tigress, of St. John's, under command of Captain Bartlett, who rescued them in latitude 53° 35′ N.

After remaining five days to finish their catch of seal, the Tigress turned to the southwest, and reached the harbour of St. John's on the 12th of May. The party had been one hundred and ninety-seven days upon the ice, and had floated southward through 24° of latitude, a direct distance of about 1,700 miles. This is at the rate of nearly nine miles a day.

But those upon the ice had fared better than those who remained upon the Polaris, for they reached home early enough in the season to report the disaster and start a relief party to the far north in search of the ship.

The Tigress was chartered for this purpose, and beat around all summer in Davis Strait and Baffin Bay, only to find that the Polaris had been abandoned, and that

the crew had escaped and been taken on board the
Ravenscraig, a Dundee whaler which had ventured to
the vicinity of Cape York.   From this, after a few days,
they were transferred to the companion vessels Arctic
and Intrepid, which were ready to return to Scotland.
The company reached New York by the steamer City of
Antwerp on the 4th of October, about five months later
than the other party.   Charles Polaris, the Eskimo baby
on the ice floe, is still an honored resident of Greenland ;
and one of the sailors of the party offered himself for
service on the Miranda last summer as she was about to
sail from St. John's for the far north.

Few things in all the world are more impressive
than this majestic belt of ice moving down the Labrador
current.   We may safely estimate that at the beginning
of summer it is one hundred miles wide and one thousand miles long.   Upon it, as already said, hundreds of
thousands of seals take refuge to rear their young,
while in their train follow the arctic bear and fox and
innumerable flocks of birds, all dependent ultimately
upon the food which the instincts of the seal enable him
to secure from the sea.

Large as is the supply, however, the number of hunters has so multiplied, and their weapons have so increased in destructiveness, that they are fast killing the
goose that lays the golden egg.   At the best, the seal
would be waging, against such odds, a losing warfare
for life.   But especially is this the case when the time
of capture involves the killing of the mother with her
young.   Already the dependence of these hardy fishermen is rapidly failing, and the late financial collapse of
Newfoundland is partially due to the poor success of her
sealers in recent years.

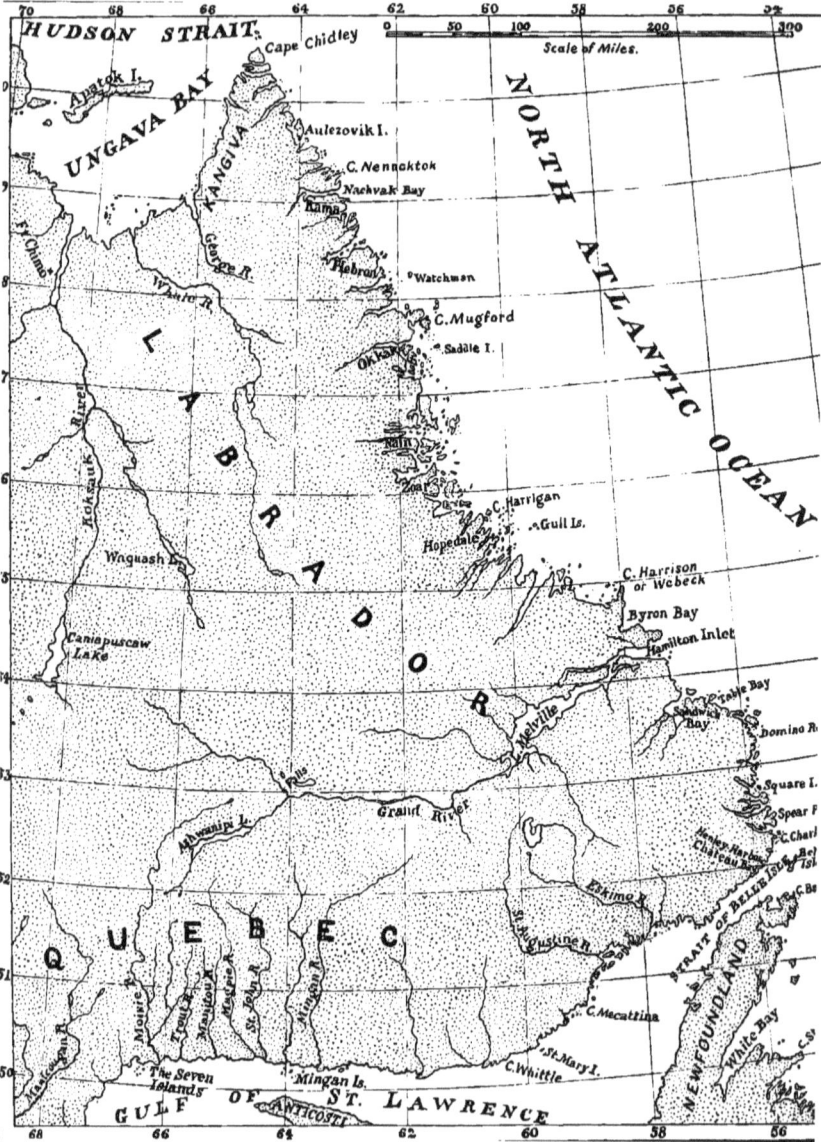

HUDSON STRAIT

Cape Chidley

Scale of Miles.
0   50   100   200   300

Akpatok I.

UNGAVA BAY

Aulezovik I.

KANGIVA

C. Nennaktok

Nachvak Bay

Kaman

NORTH ATLANTIC OCEAN

Ft. Chimo

George R.

Whale R.

Hebron

Watchman

C. Mugford

Okkak

Saddle I.

L A B R A D O R

Nain

Koksoak River

C. Harrigan

Guil Is.

Hopedale

Waquash B.

C. Harrison
or Webeck

Byron Bay

Caniapuscaw
Lake

Hamilton Inlet

Melville

Sandwich
Bay

Table Bay

Domino R.

Falls

Square I.

Manuanipi L.

Grand River

Spear P.

C. Charles

Eskimo R.

Chateau

St. Augustine R.

STRAIT OF BELLE ISLE

Red Isl.

C. Be

Q U E B E C

Moisie

C. Mecattina

NEWFOUNDLAND

White Bay

Trout

Magpie R.

Riviere R.

St. John R.

Mingan R.

St. Mary I.

C. Whittle

C. St.

Manicouagan R.

The Seven
Islands

Mingan Is.

ANTICOSTI

GULF OF ST. LAWRENCE

CHAPTER II.

AFTER our collision with the iceberg in the Strait of Belle Isle it was deemed prudent to put into the nearest port for temporary repairs. We accordingly turned northwestward to Cape Charles Harbour, on the extreme southeastern coast of Labrador, about fifteen miles distant. Our detention here, together with visits later to Henley Harbour, in the Strait of Belle Isle, and to the Punch Bowl, near Hamilton Inlet, and several days of lazy sailing in sight of the shore, gave us opportunity to see enough of the country to appreciate the broader facts which have been collected and placed on record by others.

Territorially, Labrador is a part of Canada; but so many of the inhabitants are from Newfoundland, and are in Labrador for temporary purposes, that the government of the eastern shore is turned over to the doughty little island province, which so far has refused to join the Dominion. In Labrador, as in Newfoundland, the white population is limited to the seashore, and is wholly devoted to fishing. Only about five thousand can be reckoned as permanent residents. These, in little hamlets, are scattered along the coast for several hundred miles, in conditions of life which seem to the outsider forbidding enough, but which are accept-

23

ed without complaint by the inhabitants themselves. Everywhere the aspect of the coast is barren in the

FIG. 6.—Flowing outline of the Labrador coast, with an unusually symmetrical berg in the foreground, about one hundred feet high, and two miles from shore.

extreme. No timber is in sight as one sails along the shore, and in the interior what little there is in the river valleys has small commercial value. Snow lingers throughout the entire summer in protected places, even down to the water's edge, while a long, even line of water-washed rocks bears enduring testimony to the great height and violence of the waves which roll in from the Atlantic during stormy weather.

The scanty permanent population of eastern Labrador is re-enforced during the summer by twenty-five thousand or thirty thousand fishermen from Newfoundland. For the most part these come in families; the father (and sometimes the mother) and the older children, both boys and girls, managing to combine pleasure with profit, and to make the fishing season a kind of summer vacation. The house occupied is rude and has scanty furniture, yet is not much less comfortable than one finds at many of the "Chautauqua assemblies" in the United States. Still, everything shows that the main purpose is business, and not pleasure. The girls do the cooking and keep the house, being ready, however, to devote several hours of the day to assist in cleaning the fish which the male members of the family bring ashore.

The Government of Newfoundland and the religious and charitable organizations, both of the province and of the mother country, look as well as they can after the interests of this temporary population. A line of mail steamers is maintained, running once in two or three

Fig. 7.—Storehouse at Cape Charles Harbour, with a seal skin stretched out to dry. Chapel on the hill.

weeks from St. John's as far up the coast as the ice will permit. Temporary post-offices are established at every landing place, but one will not always find them supplied with postage stamps. Usually he will pay his

postage, and trust the fisherman's daughter to purchase
the stamps when the steamer comes along. In the
winter season, when all the inlets and bays are frozen

Fig. 8.—Winter quarters in Labrador.

over and the population has shrunk to its minimum
number, the mail is carried at infrequent intervals on
dog sledges, and, strange as it may seem, is distributed
from house to house. This, however, is not so difficult
as might be supposed, since nearly everybody lives either
along the shore or a short distance back in temporary
houses made in the timber.

The isolation of many of these families is calculated
to touch the sympathies of the transient visitor. It is
not unusual to meet grown-up young people who have
never been ten miles away from the little settlement to
which they are anchored on these barren shores. Yet

upon investigation the seclusion is not so great as it seems. Trading vessels frequently call during the summer season, not only from the provinces but from all parts of Europe. From five hundred to six hundred vessels annually reach the port of Hopedale. In 1879 eight hundred vessels visited it, seventy-two lying in the harbour at one time. Packard reports that at Blanc Sablon, in Domino Harbour, there were, on the 20th of July, 1864, forty vessels awaiting the opportunity to fish as soon as the ice should clear away. As many as twelve hundred vessels sometimes visit the coast during a single season, and the exports of fish amount to two or three million dollars' worth annually.

FIG. 9.—Little chapel between the seas, Cape Charles Harbour.

On conspicuous points are built little chapels, where religious services are held regularly by laymen, and occasionally by a clergyman, who is provided with a special boat to make his long tours. Adjoining the chapel is usually a flagstaff up which a signal is run to notify the scattered population of the advent of the welcome missionary.   At Battle Harbour is maintained

FIG. 10.—Battle Harbour, the capital of Labrador.

a hospital to which the unfortunate fishermen have ready access by means of the frequent passage of vessels of various kinds up and down the coast. There is, to be sure, no telegraphic communication with the outside world, and the newspapers received are a long way behind date; yet interest in contemporary affairs is well maintained, and the political questions agitating Europe and America are everywhere intelligently discussed.

In returning from Greenland, on the 28th of August, five days after we had abandoned the Miranda amid the mists and darkness of Davis Strait, the Rigel sighted the coast of Labrador just south of Hamilton Inlet, about latitude 54°. The shortness of our supply of water and the appearance of an approaching storm led us to put into the first convenient harbour.

Here, as everywhere in the southeastern part of Labrador, the outlines upon the horizon were of the gently flowing and graceful order, which we have already remarked in Newfoundland, and which, as we shall see later, is in such striking contrast to the sky lines of the west Greenland coast. There were nowhere any sharp mountain peaks in sight, and even the numerous islands bordering the coast presented the same subdued aspect, indicating, as some would contend, recent subjection to the horizontal erosive agencies connected with a vast ice movement as distinguished from the vertical action of water erosion. This became the more evident upon reaching the shore and comparing the rocks with those in Greenland, for the geological formations are nearly identical upon the two sides of Davis Strait, showing that the diversity in contour must be attributable to the difference in the agencies which have sculptured the mountains into shape. But more of this hereafter.

Upon reaching the vicinity of the bordering islands we came in sight of numerous small boats which were out for their daily catch of fish. A schooner also hove in sight, and bore down near enough to exchange greetings and to tell us where we were. The surprise of the captain and crew of the trading vessel at the spectacle presented when our company of ninety-one persons lined up on the deck of the Rigel, filling her from stem to stern, is easier imagined than expressed. Indeed, the attempts at expression upon the part of the captain were of a character which it would hardly be permissible to put on record in a printed volume. But we learned from him that we were in latitude 53° 20', in the vicinity of the Punch Bowl, one of the snuggest of all the island harbours on the coast. When the small fishing

boats saw that we wished a pilot a lively race began, to
see which should reach us first and secure the coveted
prize.

But after the bargain was made and the pilot was on
board, the other boats gathered around us to learn our
story and to replenish our depleted larder with fish just
taken from the water. Naturally enough, also, the fishermen had wants of their own which we could supply.
One poor fellow had become nearly blind, and there was
no medical assistance within reach. He was brought by
his friends to get the advice of the skilful physicians
whom we had on board. Another came in great anxiety
for a friend who had sprained his ankle and was in
much need of some alcohol with which to bathe it. Taking compassion on him, I divided with him my ample
supply of Pond's extract; but still he begged for the
alcohol, and was so disappointed in not obtaining it that
he forgot to be thankful for what he did receive.

At length, after winding around through a tortuous
channel among the islands, we came to the entrance of
the harbour, which goes by the name of Victoria Tickle
—"tickle" being a peculiar term applied in Labrador
to many narrow and rather shallow passages between
the broader areas of water.

The Punch Bowl is well named, being a circular
body of water about a mile in diameter, with depth
enough to float the largest vessels, and good anchoring
ground. The low graceful hills surrounding it rise in
places to a height of three hundred or four hundred
feet, and are entirely without forests or trees. Abundant
vegetation, however, covers the depressions where soil has
accumulated. Just at this time the so-called "baked
apple," or cloud-berry, was ripe, and was very enticing
both in its colour and its flavour. Aside from the

whortleberries, this is almost the only edible fruit that grows in Labrador. It is of a purplish colour, in shape something like a small blackberry, and tastes more like a half-decayed than like a well-baked apple.

The day was spent by the crew in replenishing our stores of fresh water, which had been so short that for a week we had been compelled to forego the pleasure of a fresh-water bath even for our faces. By the passengers the day was spent in relaxing their limbs on shore, in recovering from seasickness, and in wandering over the lowlands and bogs in search of botanical specimens, and over the hills to learn the geology of the region. To the glacialist there was the same occasion for surprise here which had impressed us in the vicinity of St. Charles Harbour, in the fact that there were no boulders upon the hills, but that they had everywhere been swept bare and clean. Up to a height of about one hundred and eighty feet, however, there were irregularly formed terraces containing many sub-angular boulders a foot or two in diameter, witnessing to so much depression at least of the land below sea level in postglacial times. The outlook, in the light of the setting sun, from the highest hill back of the harbour was most beautiful and instructive. The island is separated by numerous silver threads of water from other islands between it and the shore—all together presenting the same flowing outline upon the horizon as that upon which we have already remarked, and merging into the scenery of the coast so gradually that it all seems to be one.

There is no permanent settlement at the Punch Bowl, but during the summer it is a favourable centre for the meeting of the fishermen and the vessel owners who are in quest of cargoes. Codfish were everywhere. The smooth granite rocks were covered far and wide with

4

them spread out to dry. All the slight elevations were capped with circular piles of fish already dried and awaiting shipment, while long rows of hogsheads were full of the fermenting livers, from which cod-liver oil with all its medical virtues is extracted.

Two vessels were lying at the dock awaiting their cargoes while we were there. Both were English ships

Fig. 11.—A house at the Punch Bowl, with the chapel on the hill.

with orders to sail with their freight to different ports of the Mediterranean—the one to Spain and the other to southern Italy—thus illustrating how the pious observances of one class of people may furnish occupation for another who are as far separated from them in their religious beliefs as they are in space. In this case the sturdy Protestant of Newfoundland is glad enough of the market provided for his wares by the fasts enjoined in distant places by the much-berated Catholic Church.

The trader at the Punch Bowl was a citizen of Newfoundland who for many years had spent his summers here. The storehouse, one or two log houses, and a small chapel constituted the settlement, but, to our surprise, everything had a holiday aspect as we sailed to our place of anchorage. Flags were flying and national colours were displayed as if it were the Queen's birthday. Coming as we did out of the darkness of Greenland, we should have been disturbed at these signs in the harbour of an English colony if there had been any impending trouble between our own and the mother country when we left home. But as the Chicago insurrection was at its height at that time, we could think of no news that should be good to them that should not also be good to us. So we took courage and hastened to inquire the cause of the rejoicing, when we learned that there was a wedding in progress. The storekeeper's daughter was married that day. The wedding, however, was not at the Punch Bowl, but in St. John's, six hundred miles away; upon which we concluded that the people in whom sentiment is so keenly alive are able to take care of themselves, and that they well may command our respect rather than elicit our sympathies. With our blessings on the far-off bride and groom, we sailed out again through the "tickle" (or rather were towed out by our dories, for there was no wind) on the following morning, and for the next three days, amid recurring calms, leisurely surveyed the coast that stretched to the south of us as far as the Strait of Belle Isle.

The geology of Labrador is comparatively simple. The prevalent rock is Laurentian gneiss, which in the southern part rises to a height of about two thousand feet at a distance of one hundred or two hundred miles from the coast. The general aspect of the interior is reported to be that of a gently undulating plateau abounding in shallow lakes connected by rather sluggish streams. To the north the plateau narrows and rises, until, seventy miles south of Cape Chidley, it is crowned with mountains which attain, according to Dr. Robert Bell, of the Canadian Geological Survey, a height of six thousand feet. The watershed runs nearly parallel with the Atlantic coast for a distance of between five hundred and six hundred miles. Scarcely anything is known about the interior, as it is exceedingly difficult of access, and there is little to attract the ordinary explorer. In general, the higher elevations appear as rounded rather than sharp peaks, indicating recent erosion by glacial agencies.

Prof. A. S. Packard cites * observations by Mr. Lieber which show that the northern mountains of Labrador, in the vicinity of Cape Chidley, have been rounded and moulded by glaciation to the height of about two thousand feet above the sea, while the higher portions of the mountains are covered with angular blocks of local origin which have been broken off by the frost and ice and moved only a little way from their native ledges. Evidently the mountain heights rose above the surface of this portion of the continental ice sheet in the Glacial period; otherwise all such loose blocks would have been inevitably borne away and deposited over the outer margin of the glaciated region. Excepting this highest part of Labrador, lying between Ungava Bay and the Atlantic coast, where the mountain tops

---

* Memoirs of the Boston Society of Natural History, vol. i, pp. 219–222; The Labrador Coast, 1891. pp. 293–296, 301.

were nunataks* during the maximum stage of glaciation, all the Labrador plateau and coast appear to have been then enveloped in ice.

Near the mouth of Hamilton Inlet there is an extensive outcrop of light-coloured gneiss of later age than that which constitutes the main portion of the interior. According to Packard, this occupies "a depression of the Laurentian rocks about one hundred and twenty-five miles long and probably twenty-five miles broad, stretching along the coast between Domino Harbour and Cape Webuc."† This rock is light coloured, only slightly schistose, and consists largely of white, granular, vitreous quartz, mingled with a small amount of hornblende and mica, but without feldspar.

This so-called Domino gneiss is accompanied with a considerable amount of coarse-grained trap which has overflowed upon it in numerous dikes. The trap rock, being of harder texture than the gneiss, presents many prominences of peculiar shape, of which Tub Island is one, its name being descriptive of its appearance. Cape North is a lofty headland of this trap with Domino gneiss underlying it. An island called Black and White, on the north side of Hamilton Inlet, consists of trap and gneiss in about equal proportions, whose colours give good warrant for the name.

One of the most remarkable remnants of these trap overflows is at Henley Harbour, in Chateau Bay, near the southeast corner of Labrador; but the ejected matter there is of a finer texture than that farther north. The most conspicuous remnant at this place is known as

---

* The Greenland name for mountain peaks which project above the surface of a glacier.

† The Labrador Coast. p. 286.

the Devil's Dining Table, and consists of a nearly circular mass of basalt, having a distinct columnar structure like that of the Giant's Causeway in Ireland. Its
surface is two hundred and fifty feet above the sea, and it
rests upon the upturned edges of the older Laurentian
gneiss, showing that it is of later age. The table consists of two distinct layers of doleritic basalt, each about
twenty-five feet in thickness. The five-sided columns

FIG. 12.—The Devil's Dining Table, Henley Harbour, Labrador.

into which it is divided are about two feet in diameter.
The flat top of the table is about five hundred feet across,
and in summer is carpeted with the bushes of the curlew
berry. Two or three large granite boulders are its sole
reminders of the Glacial period.

West of Henley Harbour there is a considerable development of Cambrian rocks consisting of red and gray

sandstones. These are nearly horizontal, and are very distinctly terraced.

The glacial phenomena of Labrador all indicate that it has been a centre from which the ice has moved outward in all directions. So far as the glacial striæ have been observed upon the eastern shore they point toward the Atlantic Ocean and Davis Strait. Hamilton Inlet was filled with an enormous glacier forty miles wide at its mouth, and extending an unknown distance into the area now occupied by water. The glacial striæ are distinct upon each side of the mouth of the inlet. On the southern coast of Labrador the eminences show that the ice movement was from the north, since their sloping or "stoss" side is in that direction; but upon the eastern coast the sloping sides are to the west and the abrupt sides to the east. In the northwestern part of Labrador the ice moved westward into Hudson Bay, from which, by a circuitous route, it flowed outward in a majestic glacier which filled Hudson Strait from side to side, being nearly one hundred miles in width.

The fullest information concerning the interior of Labrador is furnished by the report of an expedition of the Canadian Geological Survey, conducted by Mr. A. P. Low, in the years 1893 and 1894. This party entered the peninsula from the west by way of the Saguenay River, travelling in a nearly straight line to Ungava Bay, a distance of eleven hundred miles. Then, coming around by boat to Hamilton Inlet, they ascended Hamilton River to Grand Falls, and from that point explored the watershed, from which streams flow in every direction. The party came out by a southerly route, reaching the St. Lawrence opposite Anticosti Island. The valley of Hamilton River is described by them as " well wooded with white, black, and balsam spruce, larch,

balsam poplar, and white birch, much of the timber being sufficiently large to cut for commercial purposes." *

Lake Winokapau, an expansion of the river, fifty or
sixty miles below the Grand Falls, " is forty miles long,
and averages a mile and a half in width. . . . The waters
are deep to the base of the high rocky cliffs that bound
the valley on both sides. Soundings made in the centre
gave four hundred and sixteen feet. Toward its upper
end the sand brought down by the river has greatly
decreased the depth, and a number of low islands and
shoals obstruct navigation." †

In the upper part of Hamilton River valley " extensive fires during recent summers have burnt almost the
whole of the timber in the valley and on the surrounding table-land. . . . The small patches remaining show
that the trees in the valley were of fair size, while the
table-land is covered only with small black spruce and
larch." ‡

The Grand Falls, whose existence was barely known
before, were brought prominently to notice in 1892 by a
visit of Messrs. Bryant and Kenaston, and two years
later by this Canadian exploring party. Mr. Low's description is as follows: Leaving " a small lake expansion, and narrowing to less than two hundred yards in
width, the river falls two hundred feet in less than four
miles, rushing along in a continuous heavy rapid. In
the last quarter of a mile it narrows to less than one
hundred yards, as it sweeps downward with heavy
waves over a number of rocky ledges, preparatory to its
plunge of three hundred feet, as the Grand Falls, into a

* Annual Report of the Canadian Geological Survey, Part A,
vol. vii. p. 71.
† Ibid.   ‡ Ibid.

circular basin about two hundred yards wide at the head
of the cañon below. From this basin it passes out by a
channel less than fifty feet wide, at right angles to the
falls, and thus pent up in this narrow channel it rushes
on in a zigzag course from five hundred to seven hundred feet below the general level, until it issues into the
main valley below. The distance in a straight line from
the falls to the mouth of the cañon is not much over
five miles; but, owing to the crooked nature of the
cañon, the river, with a fall of over three hundred feet,
probably flows more than twice that distance before it
reaches the main valley." *

"Above the Grand Falls the character of the river
changes completely, and instead of flowing steadily in a
deep, well-defined valley, it here runs almost on a level
with the surrounding country without any valley proper,
but spread out into lake expansions and numerous channels separated by large islands, so as to occupy all the
lower lands of a wide tract of country through which it
flows. . . . The country surrounding the river is rolling,
with rounded hills seldom rising more than three hundred feet above the general surface. Between the hills
are wide valleys occupied by lakes or swampy land.
The trees are small, and black spruce predominates,
with larch, balsam, and white spruce, and a few white
birch." †

"All the lakes and rivers of the interior were found
well stocked with fish, those of the eastern watershed
especially so. During the summer of 1894 the party
lived almost exclusively on fish caught in nets or with

---

* Annual Report of the Canadian Geological Survey, Part A,
vol. vii, p. 72.

† Ibid., p. 73.

lines. The net was nightly set at random, and never failed to give a supply in the morning. Lake trout, often of large size, brook trout up to seven pounds' weight, large whitefish and pike, landlocked salmon, and two kinds of suckers, were all taken almost everywhere."

" The most important geological information obtained is the discovery of a great and hitherto unknown area of Cambrian rocks, extending north-northwest from north latitude 53° to beyond the west side of Ungava Bay. These rocks are made up of a great thickness of conglomerates, sandstones, slates, shales, and limestones, together with intrusive igneous rocks. Their chief economic value is due to the immense amount of bedded iron ore found along with them. The ores are chiefly specular and red hæmatite, together with beds of siderite or carbonate of iron. Thick beds of fine ore associated with jasper were met with in many places on both the Ungava and Hamilton Rivers, and the amount seen runs up into millions of tons. Owing to their distances from the seaboard these ores at present are of little value, but the time may come when they will add greatly to the wealth of the country.

" Frequent observations on the direction of the glacial striæ show that the ice during the glacial period flowed off in all directions from a central area south of Lake Kaniapiskau and between the headwaters of the Hamilton and East Main Rivers. Along the upper part of the East Main River the ice moved nearly due west, and it also flowed in that direction near Nichicun Lake. The striation is very indistinct, and the evidence of motion of the ice mass is not definite from here to Lake Kaniapiskau. This portion of the country is covered by immense quantities of subangular blocks and boulders of local rocks, often perched on the very summits of the rocky hills, and not uncommonly found resting on other blocks beneath, in such a position that the least movement would displace them.

" Erratics are very rare, and everything points to but a slight amount of movement of ice in this vicinity. At Lake Kaniapiskau the direction of the striæ shows the ice flow to have been toward north 60° east, while down the Ungava River it was more nearly north, corresponding with the general slope of the country. In the valley of the Hamilton River only the south side is glaciated, and the direction of the striæ follows that of the axis of the valley. On the table-land above the Grand Falls the direction of the striæ is very persistent, being constant over hill and valley, with a general direction of southeast.

" Near Lake Petitsikapau the direction quickly changes to north 50° east, apparently due to a change in the general slope. About Lake Michikamau the general direction is nearly due east. Passing southward to the Romaine River, and along it, the direction of the ice movement varies from east-southeast to southeast. On the St. John River the striæ are irregular, and mostly follow the valley.

" A marked feature of the interior is the sharp ridges of drift that lie parallel to the direction of the striæ. These ridges are chiefly composed of fine material, with well-rounded small boulders, of which a large percentage are far travelled. Where cut by the streams, these ridges sometimes show indistinct signs of stratification and may be called eskers. In detail their contour is most irregular, forming a perfect network of sharp ridges joining one another from all directions, with the material lying at very high angles impossible to obtain under water.

They greatly resemble moraines formed by the melting
of drift-laden ice at rest, and are indiscriminately scattered over the country. Terraces were observed on the
sides of the hills along both branches of the Hamilton
River. These terraces rise to over one hundred feet
above the present water level, and are so placed that they
could only be formed along the shore of a lake or lakes
formed by ice barriers.

"Almost continuous terraces were traced along the
sides of the deep valleys of the Hamilton and Ungava
Rivers from their mouths for over two hundred miles
inland. The post-glacial elevation on the Atlantic coast
of Labrador, as shown from terraces and raised beaches,
was not over two hundred feet at Hamilton Inlet, and
gradually decreases northward.

" The depth of Lake Winokapau—four hundred and
sixteen feet—would indicate that the elevation of the land
in preglacial times was much greater than at present, and
that the valley of the Hamilton River has since been
filled up with glacial drift, out of which the river is
again cutting a channel ; but owing to the less elevated
state of the land it will probably not again reach the
depth that it had previous to the glacial period." *

The inner Labrador banks are supposed by Hind to
be glacial moraines. These are situated about fifteen
miles out from the islands which border the shore, and
are covered by from one hundred and twenty to two
hundred and forty feet of water. These banks are specially noticeable opposite the fiords and bays through
which the flow of the ice was naturally concentrated.

* Annual Report of the Canadian Geological Survey, Part A,
vol. vii, pp. 78–80.

Icebergs are constantly stranded upon these shoals, forming long lines and large groups. The shoals are also the haunts of fish, which afford sustenance to the people.

North of Hamilton Inlet the coast is bordered by innumerable islands, large and small, while to the south there are very few. Prof. Hind has introduced an interesting and plausible theory to account for this striking phenomenon. Beginning a little way south of Hamilton Inlet, the coast line to the north trends rapidly westward, while south of that point for one hundred miles the direction is nearly that of the meridian. Prof. Hind supposes that the Labrador current, with its constant throng of icebergs and ice floes, and its steady westward tendency, produced by the motion of the earth, has worn away the islands off the southeast coast, and bevelled off the edge of the shore itself, while to the north, where the coast trends northwesterly, the same forces have produced a sort of eddy in which the ice is stranded and compelled to unload its burdens of drift, thus augmenting the *débris* which forms the shoals characteristic of the region. If this be the correct explanation, it is certainly one of the most interesting examples on record of the cumulative effect of a slow-working cause; for the movement of the current is scarcely two miles an hour, and even by the action of its floating ice it produces now no perceptible effects.

Labrador presents most interesting evidence of the oscillations of land which have taken place in northern latitudes since the beginning of the Glacial period. A remarkable series of raised beaches extend from Henley Harbour to Cape Chidley. At Henley Harbour a beach occurs at the foot of the Devil's Dining Table one hundred and eighty feet above the sea level. Long windrows of pebbles sweep around the island in beautiful curves, showing the gradual recession of the water. In one favoured locality from ten to fifteen such receding lines of pebbles can be seen at different levels. Around the head of Indian Harbour, on the north side of the entrance to Hamilton Inlet, Packard observed a conspicuous raised beach of wave-worn shingle, gravel, and sand, at an estimated height of two hundred feet, and a second about fifty feet above the harbour. At Hopedale, and in many other places along this coast, he also reports similar beaches. Their altitude increases northward, and at Nachvak Inlet (lat. 59°) Dr. Robert Bell observed " terraces or banks of gravel and ancient shingle . . . on either side of the inlet at various heights up to an estimated elevation of two thousand feet." Again he states that these " raised beaches show with great distinctness at an elevation of about fifteen hundred feet above the sea." *

In addition to the white population of Labrador, which is mostly confined to the portion south of Hamilton Inlet, there are about fifteen hundred Eskimos living upon the coast of the northern part, and the Indians in the interior are estimated to number about four thousand. The Indians are so isolated that they are probably the most untamed tribes upon the continent, visiting the coast only at intervals. Many things indicate that they are a waning race, and that, owing to the periodical prevalence of fires, which limit the food supply of animals, the game upon which their livelihood depends is becoming scarcer every year.

Mr. Low relates that at Fort Chimo, on Ungava Bay, a great famine prevailed among the Indians at the trading post during the winter preceding his visit in 1893,

---

* Geol. Survey of Canada, Annual Report, new series, vol. i, 1885, p. 7 DD; Bulletin, Geol. Soc. of America, vol. i, 1890, p. 308.

" whereby nearly two thirds of them, or upward of one hundred and sixty persons, died of starvation. This calamity was due to the failure of the reindeer to follow their accustomed routes of migration during the preceding autumn, when they did not cross the Koksoak River in great bands as usual. In consequence, the Indians, who depend upon the reindeer for both food and clothing, were soon reduced to starvation, and, being unable to obtain other supplies, died off by families during the winter. About twenty-five Eskimos also perished from the same cause. The surviving Indians having been in a state of constant starvation throughout the past year, and consequently being unable to trap furs and to pay their debts, were at the time of our visit in an abject state of poverty. A collection was taken up among the white people here and the officers of the steamer Eric, and sufficient was obtained to partly clothe the naked children and the widows whose husbands had died the last year. On hearing of the distress among the Indians, the Indian Department placed a sum of money at the disposal of the Hudson Bay Company this year, and a recurrence of such a disaster will be impossible in the future." *

The Eskimos have nearly all been converted to Christianity through the labours of Moravian missionaries, who were led in the work by John Christian Erhardt and four companions in 1752; but the untimely death of the leader broke up the enterprise before it was fairly started. In 1765 Jens Haven, another Moravian, visited Chateau Bay with three companions, where he found several hundred Eskimos, and remained on good terms

---

* Annual Report of the Canadian Geological Survey, Part A, vol. vii, pp. 68, 69.

with them during the summer. No permanent mission, however, was established at that time. The ability of Haven to converse with the natives in their own language, which, like Erhardt, he had learned in Greenland, was an important means of securing their good will and of promoting intercourse between them and the whites. But the bond of friendship was soon broken, and the Eskimos relapsed into a state of suspicion and warfare, in which twenty of the natives were killed in one contest. Nothing more was done until 1771, when Haven returned with a party of fifteen, and successfully established the mission at Nain, in latitude 56° 25'. In 1774 a second station was opened at Okkak, about fifty miles farther north. In 1782 other Moravians founded the station at Hopedale, about sixty miles south of Nain, and still later those at Zoar, Hebron, and Ramah, thus bringing the missions within reach of all.

In Labrador, as in Greenland, the Eskimos seem, like the Indians, to be a waning race, and few of them live to be more than fifty years old. Pulmonary diseases are extremely prevalent, and the hardships to which the men are subjected in hunting the seal make them prematurely old by the time they are forty. Those in Labrador differ little in habit and appearance from their kinsmen in Greenland, though they can not have had any communication with them for a long period—probably not for many centuries. But, as already remarked, although the early missionaries to Labrador learned the Eskimo language in Greenland, they found little difficulty in communicating their ideas to the natives on the west side of Davis Strait. The Labrador kayak, though somewhat broader and clumsier, is still very similar to that used in Greenland. The dress of the people is also in most respects very similar both in pattern and material, the principal difference being that in Labrador the blouse of the women is provided with a

FIG. 13.—A Labrador Eskimo lady in full winter dress.

pointed skirt behind, making it somewhat like a gentleman's dresscoat.

While it is no doubt true that contact with civiliza-

5

tion has been one of the chief causes of the physical
deterioration of the Eskimos in Labrador, there can be
as little doubt that the influence of the Moravian missionaries, by improving the moral condition of the people, has done much to counteract this deterioration.
Wisely the missionaries have encouraged the natives to
continue, for the most part, in their former mode of
life; but, according to all reports, the change which
has been produced in their general character is very
marked. Before coming under the influence of missionaries the Eskimos of Labrador were cruel in the extreme, so that shipwrecked sailors dreaded above all
things to fall into their hands, while cannibalism was by
no means infrequent. The change produced in their
character by the influence of the missionaries has been
no less grateful than surprising to the sailors who have
since been shipwrecked here, and thrown upon the hos-
pitality of the native population.

In judging the work of the Moravians among the
Eskimos in Labrador, one will do well to keep in mind
the just remarks of Charles Darwin when speaking of
the missionaries in Tahiti. European critics, he truly
says, are apt to compare the attainments of a newly
converted savage race not with what it was before the
advent of the missionaries, nor even with the average of
society at home, but " with the high standards of gospel
perfection. They expect the missionaries to effect what
the apostles themselves failed to do. Inasmuch as the
condition of the people falls short of this high standard,
blame is attached to the missionary, instead of credit for
that which he has effected. They forget, or will not
remember, that human sacrifices, and the power of an
idolatrous priesthood—a system of profligacy unparal-
leled in any other part of the world—infanticide, a con-

sequence of that system, bloody wars, where the con-
querors spared neither women nor children—that all
these have been abolished; and that dishonesty, intem-
perance, and licentiousness have been greatly reduced by
the introduction of Christianity. In a voyager to forget
these things is a base ingratitude; for should he chance
to be at the point of shipwreck on some unknown coast,
he will most devoutly pray that the lesson of the missionary may have extended thus far." *

The impressions of Prof. Packard, who spent some
time at Hopedale in 1864, agree so closely with my
own received from close contact with the Christian
Eskimos in Greenland, that I can do no better than
quote his language:

" The women's dress differs from that of the men in
the long tail to their jacket-like garment; some wore an
old calico dress-skirt over the original Eskimo dress—a
thin veneer of civilization typical perhaps of the education they had been receiving for the past generations,
which was not so thoroughgoing as not to leave external
traces at least of their savage antecedents. But may
this not be said of all of us?—for, only a few centuries
ago our ancestors were in a state of semi-barbarism, and
the Anglo-Saxon race can date back to Neolithic Celts
and bronze-using Aryan barbarians. However this may
be, the Eskimos at Hopedale were a well-bred, kindly,
intelligent, scrupulously honest folk, whereas their ancestors before the establishment of the Moravians on
this coast were treacherous, crafty, and murderous. To
be shipwrecked on this inhospitable coast was esteemed
a lesser evil than to fall into the hands of wandering
bands of Labrador Eskimos. The natives have evident-

---

* The Voyage of the Beagle, p. 414.

ly been well cared for by the missionaries, kept from starvation in the winter, and their lives have been made nobler and better. Even in an Eskimo tepee life has proved to be worth living." *

The diverse points of view from which different classes of people are likely to criticise each other was illustrated during our own journey by some incidents in which the Eskimo appeared to as good advantage as did the Anglo-Saxon. Among the persons on board the Miranda when she started from New York were several of the Eskimos from Labrador who had been brought to Chicago to exhibit their arts and manner of life at the Columbian Exposition of 1893; but owing to the dishonesty of one of their employers, they were left in Chicago penniless and with no means of returning to their homes. As our steamer was to touch on the coast of Labrador, friends sent them on to New York, and they were permitted to take passage with us, their wants being supplied by the generosity of Dr. Cook and various members of our party. Like most of the Eskimos of Labrador, these were Christians, and their faithfulness in observing the instructions given them by the missionaries made them an object of considerable ridicule on the part of some of the sailors. Especially were they laughed at for the reasons which they assigned for our collision with the iceberg. Unfortunately, in their estimation, we had left St. John's on Sunday, and to their simple faith our collision was a deserved punishment for breaking the Sabbath.

On the other hand, the Anglo-Saxon sailors, who prided themselves upon their superior race, had many of them been in terror during the whole of our course

---

* The Labrador Coast, p. 200.

on account of three ill omens, which did not even have
the basis of religion beneath them, but were pure superstition. First, three rats had left the ship while lying
in the harbour at New York; second, we had chosen
Friday as the day on which to depart from one of
the ports into which we had put; and, third, among
the tourists there happened to be, in the person of the
writer, a clergyman (or, as they expressively describe an
individual of his class, a "sky pilot") whose presence
on a voyage, it seems, is an omen of ill luck. The services of the regular pilot and of the "ice pilot" they
were prepared to accept, but the possible services of the
"sky pilot" they dreaded beyond measure. On the
whole, therefore, it would seem that the scruples of
the Christian Eskimo are, to say the least, as worthy
of respect as is the superstition of the ordinary British
seaman. There was certainly as much reason in the
Eskimo woman's apprehensions of evil for breaking
the Sabbath as there was in the sailor's forebodings in
view of leaving Sydney on Friday, or on account of the
instincts of the rats which chose to stay in New York
rather than risk the hazards of a voyage to Greenland.

North of Hudson Strait there is a vast region not
often visited, which is occupied by Eskimos who are
as yet not influenced by contact with Europeans.
Though the region is cold and desolate in the extreme,
it has been generally supposed that it possesses no
glaciers of very great extent; but while it is true that
the glaciers upon the west side of Baffin Bay do not
compare with those in Greenland, the region should
have credit for furnishing a small quota of icebergs to
the procession which we have described as moving
majestically southward along the Labrador coast. Dur-

ing Hall's residence in Frobisher's Bay he made observations upon an icefield in the Kingait range of mountains, from which the Grinnell glacier proceeds. This
is estimated to be fully one hundred miles long, discharging itself into the sea with a perpendicular face
one hundred feet above the water. Over this area the
line of perpetual snow comes down to within one thousand feet of the sea level, which is scarcely half the
height of the snow line in southern Greenland.

CHAPTER III.

HAVING temporarily mended the Miranda on the coast of Labrador, it was deemed prudent to return to St. John's for permanent repairs. These being completed, we started again for Greenland upon the 29th of

Fig. 14.—St. John's Harbour, Newfoundland.

July, but now we directed our course to Frederikshaab, in latitude 62°, the course being almost directly due north. Having passed at right angles through the same solemn procession of icebergs which we had viewed with such admiration two weeks before, we found our-

53

selves steaming for several hundred miles in unencumbered waters through the middle of Davis Strait.
The mountains in the vicinity of Frederikshaab were
sighted on the morning of the fourth day, August 2d;
but there lay between us and the desired harbour a belt
of floe or pan ice fifteen or twenty miles wide, and
with no openings apparent sufficient to permit our
steamer to enter it with safety. For the most part the
single pieces of ice composing this floe rose but a few
feet above the water, and were small in area when compared with those which occur in the far north. Occasionally huge icebergs, comparable in size to those seen
upon the coast of Labrador, towered in lonely grandeur
above the ice pack which here interfered with our navigation. Northward the ice extended as far as the eye
could reach, while there were occasionally narrow belts
of loose ice projecting westward from the main line like
windrows in a hayfield. These were probably distributed
by some tidal movement which was not otherwise apparent.

At this time of the year the pieces of ice forming
the floe were in a somewhat advanced stage of disin-
tegration, especially upon the borders of the belt, and
presented the most fantastic appearance imaginable.
Frequently a cake of ice of considerable extent below
the water would above the surface have the appearance
of a large cluster of mushrooms, supported on delicate
pedestals of intense blue ice merging into a basement of
green. The temperature of the water, which stood
pretty uniformly at 37° above zero, was just warm
enough to permit the waves in their continual dashing
to facilitate the melting of the lower stratum of ice
above the water line.

The coldness of the water and the great extent of the

floe ice were naturally conducive of foggy weather, which we now had for three days almost continuously. During this time but little progress could be made. At intervals of partially clearing weather the steamer would venture to move forward cautiously, but during

FIG. 15.—Floe ice west of Greenland.

a greater part of the period safety consisted in lying still. While surrounded by the fog our ears were often greeted by the ominous, dull, low murmur of the small waves which at no great distance, but out of sight, were dashing against the innumerable pieces of ice composing the floe. It was like the muffled roar of distant breakers upon a rocky coast, and, coming to us out of the mist and darkness, was calculated to affect the imagination most powerfully.

The mystery of this floe ice off the southern coast of Greenland was increased by the occasional occurrence of pieces of driftwood, some of which were from twenty to twenty-five feet in length. We were not able to

secure any specimens of these, but their story could be
readily told. They were doubtless stray representatives
of that supply of Siberian wood and timber which a
kind Providence regularly furnishes to the inhabitants
of southern Greenland to render their life endurable
and even possible. Sometimes logs sixty feet long are
drifted upon the shore. Rink reports one which yielded
between two and three cords of wood. According to
him, the pieces " are frequently twelve feet long, and a
length of thirty feet is not rare. The annual gleanings
upon the whole coast may be conjectured to be between
eighty and a hundred and twenty cords, of which
scarcely more than a tenth part passes 68° north latitude." * Much driftwood was reported by Koldewey on
the east coast in latitude 75°.

For the most part this driftwood is from coniferous
trees. Having grown upon the banks of the rivers in
far-off Siberia, these waifs were first washed downstream
in the seasons of high water, and then carried far out
into the Arctic Sea, where they were drawn into that
slow but steady current which first sets to the northward
from the northern coasts of Asia and from Spitzbergen,
and then, passing on southward, conducts the ice floes of
that region along the eastern coast of Greenland, as the
Labrador current carries southward the ice from Baffin
Bay. It is to the tender mercies of this current that
Nansen has committed himself and his companions.
Trusting to its constancy, as indicated by the few facts
at command, this heroic explorer has pushed his little
ship into the midst of the moving ice in that quarter of
the globe, and is now patiently awaiting the results, con-
fidently expecting to be carried past the north pole, and

---

* Rink's Danish Greenland, 1877, p. 91.

to be liberated at last on the southern coast of eastern Greenland.

In addition to the evidence sustaining his hope derived from the Siberian driftwood, Nansen thought he had facts of still more specific meaning in some pieces of clothing which had been lost from the unfortunate Jeannette when in 1881 it was crushed in the ice north of Siberia. After three years, what were supposed to be the same articles were found drifting past southeastern Greenland upon the floe ice of the Spitzbergen current. Let us hope that his own ship will not be crushed, that his provisions will prove adequate, and that his life and health may be spared to complete the adventurous journey.

The movement of this Spitzbergen current along the east coast of Greenland has been frequently observed, and is produced by the same general class of forces that gives constancy to the Labrador current. In 1829, when Graah was spending the winter at Frederiksdal in preparation for his expedition along the eastern coast, he seems to have known the principal facts concerning the Spitzbergen ice current, and to have speculated upon its movements about as correctly as any one could do at the present time.

"On the 25th of January, precisely the usual time of its return," he says, "the first stream of heavy drift ice, of which we had seen nothing since the month of September previous, made its appearance. The cause of its periodical departure from and return to the district of Juliana's-hope, it is not easy to determine. It is well known that the heavy drift ice usually every summer besets the southern and western coasts of Greenland, from Cape Farewell to latitude 62° or 63°, frequently to 64°, and sometimes even as high up as 67°, the latitude of Hol-

steinborg, as is said to have been the case in 1825. In
September or October, or perhaps still earlier, it disappears again, and the general opinion is that it is swept
away by the current westward toward America. No
such current, however, would seem, in fact, to exist—at
least, to the best of my knowledge, there is none such to
be met with in the district of Juliana's-hope; for which
reason I am rather inclined to attribute this regular disappearance of the ice toward the close of summer to another cause—its gradual dissolution by the heat of the
summer sun, and the sea perpetually washing over it;
the more so because detached streams of it are often seen
the whole year through, even at those seasons when the
main body of it has disappeared. But how are we to
account for the *coming* of the ice to these coasts at a certain fixed period of the year? The following appears to
me the most reasonable explanation of this phenomenon :
The ice that in January reaches these coasts is probably
part of a formation that has taken place on the east coast
of Greenland in a high northern latitude, and from which
it has probably detached itself as early as the winter
previous. It is without doubt identically the same ice
among which the Spitzbergen whalers have navigated
the summer before. By the southwesterly current, known
to prevail in these seas, it is carried down between Iceland and Greenland, to past Cape Farewell, where it encounters another current that carries it up to Davis
Straits. But as the southwesterly current here spoken
of is not accidental nor periodical, but constant (it being
the effect of the earth's revolution on its axis), and as
the polar sea contains such enormous masses of drift
ice, might we not, then, look to find Cape Farewell
always beset with ice? Yet this, as well-informed persons testify, is by no means the case, the sea around

this promontory being usually free of ice, or nearly so, from October to January. How are we to account for this? Either by supposing that the ice is broken up and dispersed by the hard southerly gales that prevail here in the autumn and winter, or that the whole mass of ice that in the spring begins its progress from between Spitzbergen and Greenland, and which reaches the latitude of Cape Farewell toward the close of summer is then already near its period of dissolution. While in the meantime this process is going on with respect to that portion of the ice that drifts toward Cape Farewell, another and considerable body of it is carried by the current in toward the east coast, where, encountering the land, it accumulates into a compact mass which only now and then yields to a strong and long-continued wind from off the shore; and which, there being here neither swell nor current to act upon it, forms with probably but little intermission a constant and impenetrable barrier along the coast." *

Singularly enough, the Spitzbergen ice current, like that of Labrador, has been made available for transportation of shipwrecked sailors through long distances. Indeed, a considerable portion of the eastern coast of Greenland, for about six hundred miles southward from latitude 68° N., was within sight of the drifting crew of the ship Hansa of the North German Exploring Expedition, who, when their ship was crushed, had sought refuge upon the same masses of ice which caused the destruction of their vessel.

The Hansa belonged to the Second German Arctic Expedition of 1869 and 1870. She was a brig of sev-

---

* Narrative of an Expedition to the East Coast of Greenland, pp. 54, 55.

enty-six tons, commanded by Captain Hegemann, with fourteen officers and men. The vessel sailed from Bremen on June 15, 1869, aiming to reach the northeastern coast of Greenland in time to do some exploring before winter set in. The Hansa was accompanied by the Germania, a screw steamer of one hundred and forty tons, commanded by Captain Koldewey, who had charge of the whole expedition. The ships planned to meet at Sabine Island (latitude 74° north); but early in September, when near the place of rendezvous, the Hansa became entangled in the ice pack and was frozen in, so that the two ships failed to meet. It soon became evident that the party would have to pass the winter in the ice, and that safety might require them to abandon the ship. A small house was built of coal bricks on the ice floe, and was made as comfortable as possible, having been provisioned for two months; but for some time the ship remained near them, secure in the ice, which was then drifting without much commotion.

From the 5th to the 14th of September they drifted seventy-two miles in a south-southwest direction. On October 18th the ice began to "thrust," and more seriously endanger the vessel. On the 19th the ship was dismantled, and as much as possible of her cargo was transferred to the ice floe upon which the house had been built. Here, buried in the accumulating snows, they passed the long winter. The coldest weather experienced was on December 18th, when the thermometer fell to 20° below zero (F.). Until January 1st little occurred to break the dull monotony of their experience; but on the 2d of January the imprisoned crew began to hear the ominous sound produced by the scraping of the floe upon the ground. They were nearing the shore or passing over shoals, and it was evident that there was imminent danger that the unstable foundations on which they had built would break up. On the 4th day of January the cake of ice upon which their house was built was diminished in size from a diameter of two miles to a diameter of one, and upon three sides of them they were but two hundred steps from the very edge, with a terrific storm raging. When the sky cleared they found themselves within sight of Capes Buchholz and Hildebrandt, and only two miles distant from them. The following extract from the captain's log gives a vivid impression of their experiences during that and several other nights:

" The weather in the past night was calm and clear. The moon shone brilliantly; the northern lights and the stars glittered upon the dead beauty of a landscape of ice and snow. Listening at night, a strong, clear tone strikes the ear, then again a sound as of some one drawing near with slow and measured steps. We listen—what is it? All still; not a breath is stirring. Once more it sounds like a lamentation or a groan. It is the ice; and now it is still, still as the grave; and from the glance of the moon the ghostly outlined coast is seen, from which the giant rocks are looking over to us. Ice, rocks, and thousands of glittering stars. O thou wonderful, ghostlike night of the north !"

On January 11th the ice upon which they were floating again split up, so that it was only a hundred and fifty feet in diameter, and the current was impelling them madly along, threatening to dash them against an iceberg ahead, which they had no means of avoiding. But fortunately the danger was passed without injury. On February 1st a fragment broke off from the main mass of ice, showing that its thickness was about thirty feet, which gave them assurance of comparative safety. On March 18th observation showed them to be in latitude 68° 2'. Thence they drifted more than six hundred miles southwesterly along and in sight of the eastern coast of Greenland. On March 29th they were in the latitude of Nubarlik, where the ice on which they were floating was pressed into a bight, and they were compelled to remain four weeks in idleness. After three weeks more they had drifted to latitude 61° 4'.

Spring was now beginning to shed its genial warmth, and, though no land was in sight, linnets and snow buntings appeared in great numbers. On May 7th open water leading toward the land appeared in latitude 61° 12', upon which, at 4 P. M., after having been upon the ice floe two hundred days, they launched their boats and took final leave of their icy foundations. On the 24th of May they reached Illuidlek Island (latitude 60° 55'), or rather the ice floe surrounding it, for it was not until June 4th that they actually reached land. On June 6th they set sail for Frederiksdal, camping at night five miles north of Cape Dalloe. Here they were greeted by the first flowers of summer—sorrel, dandelions, cinquefoil, lifting their tiny heads from every sheltered fissure which faced the sun. On the 13th of June they reached the most southern station of the Moravians, for which they had set out, where they were heartily greeted both by the natives and by their own missionary countrymen, and in due time were carried back to Europe, to electrify the world with their marvellous story.

Nansen, too, when attempting to reach the eastern coast of Greenland to begin his celebrated journey across the inland ice-sheet, was himself a prisoner amid the

FIG. 16.—A landing upon the rock, showing a trap-dike.

6

floe ice for a month.   On June 28th, in latitude 66° 24′
north, land was first sighted by the ship which was to
leave him, but it was not until the 17th of July that
he made the attempt to land.   Putting off in his small
boats, and leaving the vessel to return to Norway, he
endeavoured in vain to reach the shore, finding it impossible to do so on account of the ice.   After struggling
two days with the shifting ice-laden currents, he took
refuge upon an ice floe, and set up his tent upon it to
await the issue.   Here he was compelled to remain until
the 29th of July, when a fortunate turn in the current
enabled him to effect a landing, but not at the place for
which he had set out.   During these twelve days of
anxious experience upon the ice he had drifted two
whole degrees southward, at an average rate of ten
miles a day, corresponding very closely to that of the
Tyson party on the coast of Labrador.

When the Spitzbergen ice reaches Cape Farewell it
is forced, by the general movements occasioned by the
Gulf Stream (a branch of which runs far up into Davis
Strait), to turn northwest and hug the western shore of
southern Greenland.   It is thus that Siberian wood is
brought to supply the Greenland Eskimo with the material needed in the construction of his houses, his boats,
and his implements.   Upon this ice also are brought
the seals, which to such an extent supply him with food,
and with covering for both himself and his boat.   It
was this belt of Spitzbergen ice which our ship encountered off Frederikshaab on the 3d of August, and which
prevented our reaching the shore until the 7th.   The
Danish vessels avoid the ice by keeping about a hundred
miles south of Cape Farewell, and going northward near
the middle of Davis Strait to the latitude of Godthaab or

Sukkertoppen, when they can usually reach shore without difficulty. So continuous is this belt of ice that some of the southern settlements are rarely reached by the direct route, but vessels are compelled first to go round the northern end of the ice pack, and follow down the shore through the clear space usually existing there.

CHAPTER IV.

THE "outskirts" of Greenland, as they are called, consist of a fringe of islands, mountains, and promontories surrounding the vast ice-covered central portion, and varying in width from a mere border up to eighty miles. Upon the east side this fringe is everywhere exceedingly narrow, and affords but scanty opportunity for the maintenance of life of any kind. Upon the west side, below the seventy-third parallel, it has an average of about fifty miles in width, and extends with little interruption from Cape Farewell to Melville Bay, a distance of something over one thousand miles. Everywhere this mountainous belt is penetrated by deep fiords which reach to the inland ice, and are terminated by the perpendicular fronts of huge glaciers; while in the vicinity of Ivigtut (latitude 61°) and Frederikshaab (latitude 62° 45') the ice comes down in broad projections close to the sea margin.

The seaward aspect of the west Greenland coast is stern and forbidding in the extreme. The serrate edge of the long mountain chain does not, however, rise to any great height, being rarely over two thousand or three thousand feet above the level of the sea, with occasional peaks running up to four thousand feet. It is only in a limited area north of Disco Island that the mountains rise to a height of seven thousand feet.

66

R A F F I N

Sukkertoppen

S. S. Miranda went
aground, Aug. 5, 1894.

South Isortok F

Sangu

As first seen from a distance of forty or fifty miles, Greenland seems anything but a justification of its

FIG. 17.—View from the harbour of Sukkertoppen, looking east.

name, for even the "outskirts," which are supposed to be free from ice, are so only in part. Local glaciers, which would be objects of great attraction in Switzer-

land or Norway, mark the summits of many of the
promontories, and from a distance form a conspicuous
element of the scenery. As one approaches nearer,
these lingering ice masses upon the summits become
hidden from view behind the projecting precipices and
steep slopes of the partially submerged mountain range.
But still there is little to justify the name of Greenland.
No forests or shrubs and no running vines relieve the
sternness of the rocky surfaces. Even the lichens and
mosses, by their sombre hue, intensify the barrenness of
the scene. Upon penetrating the fiords, however, a partial change takes place. A few miles back from the
border in southern Greenland there are numerous small
expanses of pasture land and a few limited areas covered
with stunted shrubs and dwarf trees.

According to Rink, "the largest and tallest birch
tree" in Greenland is fourteen feet high, but this height
has been attained only through the protection of two
huge boulders between which it is so fortunate as to be
sheltered on either side. Willows and alders frequently
grow in the south to a height of from five to eight feet,
while juniper bushes sometimes attain a thickness of
five or six inches, but these are merely creeping shrubs
spreading out over the tops of the stones and rocks and
attaining to no height. In the sheltered places numerous brilliantly coloured flowering plants abound, of
which rhododendron, epilobium, the bluebell, the arnica, and the buttercup are prominent. But these are
not conspicuous in the general survey of the country.

A favoured spot, which Rink has called the " Greenland Eden," occurs between Lichtenau and Frederiksdal
(latitude 60°). The place is about twenty miles back
from the sea, a little to the east of the middle portion of
Tasermiut Fiord. Here, on passing out of the fiord, up

the rapids of a small stream, one finds himself in an amphitheatre surrounded by sheltering mountains three thousand feet high, where vegetation flourishes as nowhere else in the outskirts, and where the largest tree, already spoken of, had found opportunity for growth. The enterprise and sagacity of the early Norse settlers is shown in the fact that they had discovered this nook and occupied it, as is made known by the extensive ruins of several stone buildings. Nearly all of the early Norse settlements were in such sheltered places back some distance from the sea margin and south of the sixty-fifth parallel.

The greenness of these sequestered nooks furnishes some justification for the name of the land; for, after beating about amid the floes of Spitzbergen ice which encircle the southern portion of the country, and after having penetrated the various fiords intersecting the frowning seaward wall, the little green patches which at length greeted the adventurers well may have deeply impressed their minds. It is more probable, however, that the name had a different origin, being chosen to promote a land speculation, as is recorded in the history of Eric the Red. " Let us call the name of the land Greenland," he is reported to have said, " because people will sooner be induced to go thither in case it has a good name." So successful is this scheme of the crafty adventurer said to have been that twenty-five shiploads of fortune-seekers followed him from Iceland, their less attractively named but far more hospitable native country.

Twenty or twenty-five miles back from the shore, in southern Greenland, there is everywhere a marked amelioration of climatic conditions, and much less prevalence of foggy weather than on the coast. The favourite reindeer haunts are in this comparatively sheltered belt lying adjacent to the edge of the inland ice. In some years as many as fifty thousand reindeer have been killed by the natives in these hunting grounds. The problem of how this useful animal came to be dispersed through so many degrees of latitude and along a shore that is frequently interrupted by impassable fiords and icefields will be considered in a later chapter.

The division between northern and southern Greenland on the west coast is fixed at Nagsutok Fiord (latitude 67° 40′). South of this the bays and inlets do not freeze over in winter sufficiently to admit of being traversed with dogs and sledges, so that the colonies there are much more isolated than in north Greenland, where natural highways over the frozen water invite much travel in the winter season, and where the brilliant moon and the flaming aurora vie with each other in dispelling the gloom of the long arctic night.

The geological features of the western coast are remarkably uniform, except in the vicinity of Disco Bay. The rocks, like those of Labrador, consist largely of gneiss and granite, intersected with numerous dikes of eruptive material; but, so far as observed, there are no extensive areas of volcanic rock in southern Greenland. Nor have any mines of the precious metals been found. At Ivigtut (latitude 61° 10′), cryolite (a fluoride of sodium and aluminium) has been discovered in such extent and purity that it has been profitable to mine it for the manufacture of soda and alumina, the latter being of a quality much desired in the art of dyeing. There is no other place in the world where this mineral is found in large quantities. The vein here occurs between walls of gneiss and is three hundred feet wide.

First and last our view of the Greenland coast ex-

tended from Frederikshaab to the vicinity of Holsteinborg, a distance of about three hundred miles; while
from Sukkertoppen (latitude 65° 25') we were able to
make extensive tours into the interior through the fiords
and along the channels.   A brief account of two excursions will assist in bringing the general features of the
country more vividly to view.

Having arrived at Sukkertoppen on the morning of

FIG. 18. –Scene on the way to Isortok Fiord.

the 7th of August, it was soon ascertained that we were
to remain two days before setting out for the regions
farther north.   A party of ten or twelve was therefore
organized to visit the glaciers and reindeer pastures
about twenty-five miles up South Isortok Fiord, which
lay directly to the east of our anchoring place.   Isortok
means "having muddy water," and hence is descriptive
of those fiords which are discoloured by subglacial
streams of considerable size.   On this account the name

is found applied to more than one fiord along the western coast. This one reaches the sea in latitude 65° 25',
and penetrates the border to a distance of about fifty
miles, where it meets the inland ice; but branches from
the ice come down to it from a projection on the north,
about twenty-five miles from Sukkertoppen.

The scenes connected with our setting out upon the
excursion were novel and exciting in the extreme. We
were loaded into three of the ship's boats (one large
boat and two dories) with a trusty Eskimo guide in each,
and were provided with camping outfit and a limited
supply of provisions. It was past the middle of the
afternoon before we were well under way; but the sky
was clear, and as evening twilight lingered until the
break of the following dawn we had no fear of being
benighted.

In response to the lusty stroke of our oarsmen we
were soon out into the middle of the shallow bay, where
the mountains rose in picturesque forms both before
us and behind. Behind us they continued to rise
higher and higher above the lower but nearer prominences, until the serrate edge of the central axis came
full in view on the glowing western horizon, where
the contour of jagged edges was so striking that it
would be difficult to find anything anywhere in the
world to match it. It reminded me of nothing else so
much as of the sky line of the Teton Mountains as I
saw it a few years ago from Jackson's Lake at their
eastern foot, looking toward the setting sun; but here
the beauty was enhanced by the extensive and variegated expanse of water which was spread everywhere
around us, extending its arms into countless recesses of
the islands, or stretching out through illimitable vistas
into the retreating fiords. Before us the mountains of

the mainland gradually sank behind the innumerable islands which we approached and among which we slowly threaded our way. But now and again they would burst forth in new glory as some special island point was rounded and we came out for a little while into an open stretch of water. In lee of the islands the waters were as peaceful as on a landlocked lake, but when the broader passages were reached the swells of the neighbouring ocean tossed our boat in a manner well calculated to arouse the fear of a landsman, while all were inclined to give wide berth to the breakers which dashed against the windward shores or marked the shallow reefs whose backs were almost bare at certain stages of the tide. Here and there a piece of driftwood had been safely hauled ashore by some native and placed above the reach of the waves, to await the convenience of the finder. So sacred is the right to this kind of property that no one thinks of appropriating what another has discovered. Scarcely anything else is invested with such well-recognised property rights.

The sun went down long before we reached our objective point; but we rowed on in the brilliant twilight until about eleven o'clock, when it was decided to pull in to shore and encamp for the night. We were now fairly within the fiord. In rounding the point at the entrance several deserted igloos indicated the attractiveness of the neighbourhood for temporary residence, but only the Eskimo knew where to find safety and comfort for the night. Passing one or two places which looked attractive enough to inexperienced eyes in the twilight, we at length rounded a low projecting rock, and entered a sheltered cove, where we could draw our boats far out upon a sandy beach beyond the reach of the rising tide. A few steps away there was a level plat of moss and

matted running vines which made the softest imaginable
bed on which to spread our blankets and stretch our
weary limbs, for each of us had had his turn at the
oars. The tent could shelter but half the company, but
the sky was so clear and the weather so moderate that
the rest could sleep in their bags in the open air without

Fig. 19.--First camp on the fiord.

discomfort. It was so light that we could read the time
on our watch dials in the tent all night long.

Although daylight appeared behind the mountains
of the eastern side of the fiord long before it had disappeared in the west, the sun did not surmount the line
of peaks until the forenoon was well begun. Folding
our tent and eating a hasty breakfast, we set out at an
early hour to complete what seemed to be the short pull
to the mountain (Nukagpiak) which had so long en-

tranced our vision, its snowy sides rising like a fairy object above the nearer but lower peaks, full in view all the way from Sukkertoppen. It was now high tide, and the water was almost without a ripple. Everything, therefore, promised a speedy completion of our journey. The walls of the fiord rose in increasing grandeur on either side, and the local glaciers and snowfields took on ever-changing and fantastic forms upon the flanks of the mountains as we shifted our position in the channel. The width of the fiord was apparently a little over a mile, widening out into broader expanses at infrequent intervals. Numerous flocks of gulls flew over our heads, and the reports of the shots fired at them from our guns reverberated from side to side in a most impressive manner, revealing, by the length of time which separated the echoes, both the width of the channel and our relative distance from each mountain side.

As at one place we rounded a promontory we came suddenly in sight of a party of natives in a boat, who were somewhat alarmed at our approach and at the reckless firing of our guns ; but their fears were soon allayed, and we approached near enough to them to find that they were a family from Sukkertoppen who had been spending the summer at a neighbouring camping place, where game and fish were plenty, and now were returning home to make preparations for the winter. The women were at the oars. As the water was smooth the kayak was resting on the rear of the boat, and everything betokened the pleasure which they derived from the beauty of the scene and the leisurely rate at which they were permitted to proceed. From them we purchased a supply of freshly caught salmon trout, and then pulled away with all our might to reach the foot of the mountain peak whose vision had been so long tantalizing us. But the tide had turned, and though the surface was smooth a swift current was setting outward, to discharge the vast amount of water which the rising tide had pushed into the upper twenty-five miles of the fiord. On this account progress was slow, and at times we were scarcely able to make any at all. The mountain, however, did gradually grow nearer to us, and the hanging glaciers from the projecting plateau which buttressed its southwestern side smiled down upon us, as though they were the most innocent objects in the world; whereas they were in reality the most terrible, being liable to break off at any moment and rush down the steep sides in swift avalanches to the water.

At length, but not until the forenoon was nearly passed, we attained the cove for which we had been aiming, and, pulling our boats above the reach of tide, we hastened to explore the strange scenes of the vicinity. A rich growth of grass covered the rocks in the narrow inclosure and partially disguised their ruggedness. The milky current of a brook which issued from the foot of a great glacier a few hundred yards away rushed madly over the boulders which lay in the bed of the stream. To the south of us the northern face of a portion of the mountain range frowned upon us with its jagged peaks, its numerous hanging glaciers, and its bare perpendicular walls. North of us was the sloping sunward flank of Nukagpiak, covered with verdure and brilliant flowers, where reindeer might bask in the sunshine and feed to their hearts' content during the long summer days.

For two or three hours we wandered over this Elysian field, plucking its flowers, dancing in delight on its thickly carpeted, quaking bogs, and clambering to its various points of lookout, from which the eye could at one sweep take in the whole view from the island-dotted coast to the smooth white outline of the vast inland ice-fields. The glacier whose subglacial stream entered the fiord at our landing place came down to within about three hundred feet of the sea level. It did not present a perpendicular face, but ended in a very steep slope—too steep to permit a direct ascent of its surface. To get upon the ice it was necessary to clamber along the lateral moraine to a height of about one thousand feet before the slope became sufficiently gentle to render it safe to venture upon it. Here it was not far from a mile in width, growing wider higher up, until it merged into a large snowfield which covered the extensive plateau of which Nukagpiak is the culminating peak. This glacier is isolated from the main ice-sheet, but upon the north side of the fiord vast snowfields continuous with those of the interior are visible. East of Nukagpiak, upon both sides of the fiord, there are large areas from which the snow melts off in summer, and which furnish pasturage for a considerable number of reindeer.

We had heard much about the Greenland mosquito, but here we met the creatures themselves and both saw and felt them in all their glory. They came out in swarms from every tuft of grass and bunch of flowers, and although less voracious than those in temperate climates, they made up in power to produce discomfort by the infinitude of their number, which rendered it almost impossible for us to rest a moment. The various members of our party, as they were scattered over the mountain side, looked like moving windmills, so vigorous and constant were the motions necessary to get even partial deliverance from these pests. The sight of one of our number who attempted to bathe is never to be forgotten. Before he could reach the water's edge the

Fig. 20.—Ikamiut, looking north. An igloo to the right.

mosquitoes had gathered upon his naked skin, until at a distance he looked like a hairy ape; and when he plunged for relief into the cold water they hovered around, as if well knowing that it was too chilly for him to endure it long, ready to light in clouds upon him as he rose from its depths. It is fair to say, however, that this was the only occasion upon which we were troubled with the pests. In the much longer excursion taken at a later time the mosquito netting with which we provided ourselves was a useless article. Either it was too late in the season or the weather was too cold and stormy, or for some other reason, they did not visit Ikamiut, where we were in camp during the next two weeks.

It was late in the afternoon when we set out upon our return from Nukagpiak, and we were doubtful whether we should reach Sukkertoppen that night. But our provisions were nearly exhausted, though we had supplemented them by a hearty meal upon the fish purchased in the morning. These we had boiled until they were tender and had eaten without seasoning. By general consent, however, it was agreed that, even so, we had never had a more luxurious repast.

The tide was now low, and we found that our boats were left a long way from the margin of the fiord and with anything but a smooth channel leading to it. The bouldery bed of the mountain torrent rendered it exceedingly difficult to haul the larger boat down to the water without injuring it; but having accomplished this task, we made all haste to get well on our way before the tide should rise and delay us by its incoming, as it had done in the morning by its outgoing current. The mosquitoes proved faithful in their friendship and did not forsake us until we were some miles on our way. Such progress was made in the early part of the evening

7

that it was thought best to make no stop, and so we had
the pleasure of again wending our course through the
islands of the bay in the middle of the night, when the
picturesque outline of the mountain peaks around the
northern horizon was again all aglow with the twilight of
the midnight sun. The Miranda was reached about two
o'clock in the morning, to find that the captain had
determined to start northward as soon as possible.

After running upon the reef outside the harbour
and returning again to our anchoring place, it was decided that it would be necessary to remain ten days, at
least, at Sukkertoppen, while the kayakers went up the
coast to Holsteinborg in search of assistance. To relieve
our friends in case we were compelled to spend the winter in Greenland, all wrote home letters and despatched
them by kayaks to Ivigtut, about three hundred miles to
the south, hoping that they would reach that point before the last ship left for Denmark. The faithfulness
of the messengers is witnessed to by the fact that in due
time the letters all reached our friends, although two
months later than our own arrival home.

As there was no time to be lost, another camping
party was immediately organized, to be absent from the
ship ten days to do what it could in exploring the edge
of the inland ice, which comes down into the fiord setting back from Ikamiut, twenty miles north of Sukkertoppen.

Again the expedition started in the middle of the
afternoon. One large boat and two dories were required
to carry us and our equipment, while three kayakers accompanied us for our protection and assistance. The
swells which came in from the southwest were long and
high until we reached the lee of a line of islands, in which
our guides were careful to keep us as much as possible.

FIG. 21.—Sermersut, four thousand feet high, with the village in the foreground.

We were now passing through one of those long channels between the picturesque mountains whose vista had
so delighted us on previous days. In due time great
glaciers began to look down upon us from the mountain heights to the east ; but they paused in their
course long before reaching the water level. Near here
a broad opening to the ocean displayed itself between
the islands of Sukkertoppen and Sermersut, and permitted the swells from two directions to toss us upon their
capricious crests. A hard pull now across the mouth of
Ikamiut Fiord brought us late at night, but still amid
the splendour of the arctic twilight, to the settlement on the point of the promontory at the northern
side of the fiord, where it joins the open channel east
of the large island of Sermersut. To our unpractised
eyes there were no signs of human habitation near, but
on rounding a low projection of rocks our ears were
greeted with the indescribable jargon of a strange dia-
lect proceeding from the throats of twenty-five or thirty
Eskimos, young and old, who had crawled out from the
most miserable-looking human habitations that it is possible to imagine. But they were friendly voices, and we
did not scorn the help rendered us in unloading our
boats and hauling them to a place of safety, nor the
advice given us as to the most suitable camping place.

In the morning we took more careful note of our
situation, and of the condition of the people who were
to be our neighbours. Across the channel, at a distance
of about three miles, rose the picturesque eastern face
of Sermersut Island to a height of something over four
thousand feet, showing clearly the westerly dip of the
strata, and concealing the vast icefields which cover
the northwestern slope of the island. Amid the fogs
and rains and snows of the next two weeks this moun-

tain outline was destined to fix itself in our memories in innumerable aspects which could never be forgotten. The interest of the scene was enhanced by the squalor of the igloos of the Eskimo in the foreground. Of these there were only three, occupied by twenty-five people. They consisted simply of walls of stone and turf about twenty feet square and three and a half feet

Fig. 22.—A typical igloo.

high, covered over with a slightly conical turf roof, through which, in one or two of the cases, a stovepipe protruded, for use on the occasions when a fire was built in the sheet-iron cylinder which served for a stove inside; but the turf is usually so wet that most of the time a fire is entirely out of the question.

The squalid condition of the igloos was partly due to a flood which had swept over the village in the spring. How a flood could have risen in such a situation it was difficult for us to see, but the fact had to be accepted, for the ruins of an igloo in which two or three of the inmates were drowned was a mute but constant witness to the sad event, and the vivid memories of the poor survivors enabled them to make us understand the story even when told in an unknown language, so expressive were their gestures and pantomimes. In August a small stream of pure water from the melting masses of snow which still lingered in the low, rocky mountain rising above the settlement on the east rushed merrily down past the place, furnishing an unfailing supply for summer use. But it seems that when the deep snows were rapidly melting in the spring this channel became so clogged with masses of snow and ice that the water deserted its natural bed, and in a manner which seemed incredible rushed directly across the neck of the low peninsula to the opposite side from that of the natural depression. The possibility of such a destructive flood in such a situation gave us an idea of the accumulation of snow in the winter which we could not otherwise have obtained. It would seem that during most of the winter the snow is so deep that the igloos entirely disappear beneath it. The entrances to them must then have looked still more like burrows than in the summer.

Notwithstanding this forbidding exterior of the village, we found the inhabitants the best of neighbours, faithfully practising both the outward observances and the moralities of the Christian religion which had been taught them by their Danish protectors. One of their number acted as catechist, and conducted regular Sunday services in the largest of the igloos, and all the adults could read and write, though their outward garb was the traditional one which had characterized the people from the earliest times. In another chapter will be found a description both of them and of the Sunday service which it was our privilege here to attend.

The fiord which we planned to explore extends eight miles inland from the point on which we were encamped, and is from two to three miles wide, though from the clearness of the atmosphere it was difficult to make either of these distances seem half so great. The solemn grandeur of the scenery exceeded anything which it had been our privilege elsewhere to behold. The mountains rose on either side to a height of something more than four thousand feet, which, indeed, is not so high as may be found in many other parts of the world; but the interest is not exhausted in the consideration of any single feature of the scene. Opposite to the entrance of the fiord was the picturesque outline of the peaks capping the island of Sermersut, which alone separated us from the waters of the ocean, while at the head of the fiord a broad projection from the inland ice-sheet came down on both sides of a high mountain peak to the water's level and broke off into icebergs, which were slowly floating outward toward the sea.

Nothing could be more striking than the contrast between the opposite sides of the fiord. The flanks of the mountains on the south side, facing the north, were deeply covered with snowfields and furrowed with glaciers. Above the snowfields a series of sharp needle-like peaks projected just enough to give savage variety to the scene. On this flank the local glaciers presented an object lesson most perfect of its kind. A series of glaciers approached the water level at the base of the mountain to distances approximately proportionate to that separating them from the ice front at the head of the fiord. Near the entrance was one coming down to within about one thousand feet of the water level.

Farther east was one reaching to within about five hundred feet of the level. Farther east, still another came to within about three hundred feet; while still nearer and within about half a mile of the main projection of the ice front was one extending to the water's edge, and sending off miniature icebergs to aid in cumbering the waters of the fiord.

A singular feature of all these glaciers on the slope of the mountain on the south side is that when viewed from the head of the fiord they seem to be much thicker near the bottom of the mountain than they are at the higher part of their levels. This appears in the photograph taken from a point two or three miles back from the front of the glacier which comes in at the head of the fiord. From all these glaciers fragments were occasionally breaking off and falling into the water with their customary loud reports, which echoed from cliff to cliff, like the bombardment of a stronghold by modern artillery.

The mountains rising from the north side of the fiord and facing the south presented a most striking contrast to those on the south side. They were of as great height, and were equally picturesque in their outline, but from them the direct rays of the summer's sun had caused the glaciers to melt and green verdure to spring up wherever any soil was preserved. There were no trees, or shrubs even, but at frequent intervals the rugged ribs of rock which form the larger part of the mountain side were interrupted by the richest imaginable masses of green moss and matted blaeberry vines clinging to the sides where to the eye it would seem that everything living must have been swept downward by its own weight. On nearer approach every nook and corner was found to be full of most brilliant and beautiful flowers, which had been waiting through countless generations for some visitor to appreciate their significance and beauty.

About two thirds of the distance up the fiord there was a favourite haunt of various kinds of birds. At the time of our visit kittiwakes were there in countless numbers. The perpendicular precipices, for a mile or more in length and more than a thousand feet in height, were completely covered with their nests wherever there was a crag upon which they could be built. Indeed, the face of the cliffs was white with these birds as they struggled with each other to secure places for temporary rest, while the neighbouring waters were covered with those who were seeking for food or were enjoying the luxury of a cold bath. The firing of our guns would be the signal for the whole colony to rise into the air, when it would seem as if a cloud had suddenly cut us off from the sunlight, while the sound of their strange voices, whose note is imitated in their name, filled the air and completed a scene that can not be equalled in interest outside of Greenland.

With all the apparent unpropitiousness of Nature in this place, there is much to attract the natives, who know how to utilize its advantages. The three requisites for the existence and comfort of the native Greenlander are here easily obtained. Fish of various kinds come in their season, and literally wait to be caught in the vicinity of convenient projecting rocks. When we were there it was the season of cod, which could at any time be obtained almost without effort. Native boys, with the most primitive hooks and lines and almost no bait at all, would stand on a rocky projection and draw out from the water enough for a meal in an incredibly short time. Numerous piles of fish which had been

FIG. 23.—Ikamiut Fiord, looking east. The main glacier eight miles distant. The island is two miles this side.

dried and stored up for later use betokened the liberal precautions which Providence has taken to secure sustenance for the people.

The birds of which we have spoken, with two or three other kinds (the ptarmigan, the auk, and the cider duck), are easily caught and are most serviceable in various ways. A kayaker can go out at any time during the nesting season and load his boat without the aid of firearms. With his noiseless kayak he can approach near enough to the flocks as they are resting on the water to secure any number with his primitive spear pointed with sharpened bone. Indeed, he is rather more sure of his game with his spear than he would be with a gun, for the noise of firearms frightens the whole flock, whereas with his spear and kayak he can steal upon them almost unperceived. Large numbers, also, can be obtained from the rocks within reach of the kayaker from his boat. The eggs likewise form no insignificant addition to the natives' larder, while the skins furnish them with the warmest of clothing. These are tanned with the down and feathers on and sewed together into undergarments or made into quilts with which to defy the rigour of even an arctic winter.

In due season, also, the various kinds of seal visit these waters, and supply the native hunter with material for covering his kayak and umiak, and for making his waterproof boots and trousers, and with abundant fat for his lamp and for his own stomach, made voracious by his exposure to the keen storms of winter. Altogether it is not surprising that the natives look upon themselves as the special favourites of Providence. At any rate, they seem to receive their gifts more directly from Nature than do the inhabitants of the temperate zones. But this very circumstance leads them to limit

their efforts to a narrow range of occupations, and
stands in the way of their attaining any high stage of
civilization.

Two excursions to the head of the fiord were specially notable, both for their results and for the occasion
they furnished to exhibit the characteristics of the
Eskimo. On the first clear day which offered we set
out on the smooth water of the fiord at full tide and
pulled with all our might to reach the island which lies
about two miles below the ice front of the main glacier.
Five were in our dory, two of whom were natives, one
being the catechist. Another boat similarly equipped
accompanied us, while two or three kayaks went along
of their own accord to collect a supply of birds from the
loomery which we were to pass. The island was reached
in due time, though it was a longer pull than we supposed it would be from the appearance in the clear
atmosphere after a storm. From this island as a centre,
the full beauty and impressiveness of the scene could be
taken in as from no other place. The entire face of the
front of the great glacier, two and a half miles wide,
was exposed from this point, while all the hanging glaciers to the south were within near range of our vision,
and were from time to time favouring us with a display
of their power by sending avalanches of ice down the
mountain side. After photographing the scenes from
every point of view which this enchanting spot presented, we entered our boat to proceed to the vicinity of
the ice front.

. But as soon as our Eskimos perceived that we were
to go farther in that direction they both struck, and
not only declined to use the oars themselves, but refused
to allow us to use them to go in that direction. Their
faces vividly showed the real terror they were in, which

was so great that for a time we thought our plan would
be completely frustrated. But one of the party, with
the little native language at his command, succeeded in
persuading the catechist that I was a very great and
good man, when suddenly all fears were dispelled and
both the natives set to work with right good will, and
we were soon landed between the subglacial stream issuing from the south side of the great glacier and the
upper one of the hanging glaciers which come down
from the mountain to the water's edge. It was probably the first time that any of the natives had ventured
so far toward the ice front, for they have a deadly terror of it. And, indeed, why should they not have?
for there is nothing to induce them to venture so far,
since they seem to have absolutely no scientific curiosity,
and there is no game to be pursued upon the surface of
the ice. It is little wonder, therefore, that the numerous ice falls and the resounding detonations accompanying both them and the formation of crevasses fill
their minds with dread of the mysterious powers here at
work with such mighty effect.

At a later time we proceeded directly to the same
landing place below the southwestern corner of the glacier, and had no difficulty in prevailing upon our native
helpers to venture to the spot; but upon our deciding
to land above the subglacial stream, a little nearer the
glacier, there was the same display of terror as before.
I had to jump out in my high rubber boots almost to
my waist in the ice-cold water, seize the rope, and pull
the boat ashore against their most vigorous efforts to
push it off with the oars. But when they saw that I
had landed without any convulsion of Nature following,
their fear was allayed, and all hands took hold to draw
the boat into a place of safety.

This, however, was no easy task, for it was evident
from the pieces of ice stranded on the beach that the
tide rose very high. This being the case, we did not
object to having two or three of the natives remain by
the boats while we spent the day upon the glacier. And
it was well that they did stay by the stuff, for when we
returned at the close of the day the water had risen so
as to cover the vast sand bar over which we had hauled
our boats and had invaded a considerable portion of the
moraine to which we had taken them for security; but
our dusky companions had been faithful to their trust,
and had patiently awaited our return and kept the
boats in a place of safety. They had also learned that
their fears of the ice when properly approached were
groundless.

For half a mile along the southwest corner of the
glacier the approach to the interior ice is up a gentle
slope, which is deeply covered with morainic material.
This can be approached at any time; but, like the natives, we felt like giving a wide berth to the hanging
glaciers on the south side of the fiord, and to the two
miles and a half of perpendicular ice front which looked
down upon the calm water at its head.

The day upon the glacier was exhilarating in the
extreme. After clambering over the crevasses and pinnacles of ice which obstructed our course for the first
half mile, we saw a clear way before us between two
medial moraines which came down from a high nunatak in the distance. While crossing one of these moraines, picking our way between its vast piles of stones,
the two Eskimos who had accompanied us thus far
began to lose their courage, and in the true native style
attempted to disguise their real state of mind by calling
attention to their boots, saying that they were "no

FIG. 24.—Ikamiut Fiord, looking south. Showing local glaciers.

good," every once in a while uttering this ejaculation and pointing to their upturned soles with a despondent look. Of course we humoured them, and permitted them to sit down with some of our superfluous luggage to guard. Here they remained all day long, apparently not having stirred from their tracks until we hailed them on our return.

We followed up the vast glacier to the *nunatak*, which proved to be fully seven miles back from the front and to about equally divide the vast ice streams which poured down on either side of it. The total width of the glacier we estimated to be here six or seven miles, and at the base of the nunatak we were not far from two thousand feet above sea level. Eastward there was nothing but the horizon to obstruct our view. We were looking out upon the same snowfields which had greeted our vision from Nukagpiak two weeks before, only now we were on the field itself. Then we had viewed it from the side, at right angles to our present vision. The imagination now came in, with its subtle power, to intensify the interest of the occasion. With the mind's eye there was nothing to hinder our looking across the whole vast waste of perpetual snow stretching to the east coast of Greenland. This was verily a part of the inland ice.

Nor was the interest of the backward glance much less impressive. The glacier at the head of Ikamiut Fiord was only half of what was within our vision. The mountain upon the south side, whose hanging glaciers had so enchanted our vision from our camping place, divided the glacier we were exploring into two nearly equal portions. One half was pouring into the fiord on the south, through whose long vista we could distinctly see the distant islands in the bay of Sukkertoppen. At various distances along this fiord icebergs glittered in the light of the declining sun, showing that the ice front at the head of that fiord was similar to that in the one which we had more particularly investigated.

As before remarked, these glaciers on the south side were all of them thicker near the base of the mountain than in their higher levels. Indeed, they seemed to run down like cold tar and to thicken at the base as a stiff semi-fluid would under the action of gravity. Usually the more rapid melting at lower levels causes the glacier to thin out near the foot, but here the temperature in the shade is so near the freezing point that the ice melts about as fast near the upper portions of the glaciers as it does at the base.

Another phenomenon illustrating the nature of the movement going on in great glaciers was seen here to special advantage. Where the great ice-sheet abutted against the mountain which divided its front into two portions it was pushed up by the momentum of the movement so as to be two or three hundred feet higher at the base of the mountain than it was a mile back. Indeed, a half mile or so back there was a distinct depression in the glacier with the ice higher all around it. It was just such a depression as is made where a current of water is obstructed by some obstacle; the current pushes some distance up the obstruction, and then breaks over the sides to go around it; but ice, being much less fluid than water, moves off in larger swells and more gradual curves.

But even an arctic afternoon has its close. With regret we sought our boats and set out on the return, to go again through the magnificent panorama of the morning. The day had been one never to be forgotten. With its pure air making everything clear within the

8

range of vision; with the consciousness that you are treading where other human feet have probably never trod, and are looking on scenes that few, if any, others will ever see; amid a solitude that is unbroken by living objects except here and there a passing bird or a wary fox, whose tracks surprised us on the newly fallen snow; with gurgling streams of purest water from the melting ice all about us hastening in channels of deepest blue to plunge at last with deafening roar into some mysterious *moulin*, the senses were overburdened with material for the imagination to seize upon and work up into pictures of scientific form and poetic fancy. We tried in vain to answer the question which involuntarily arose, Why is there so much waste of beauty and grandeur so far beyond the reach of ordinary mortals?

The Danish artist A. Riis Cartensen enjoyed the privilege of joining Captain Jensen's surveying party, which spent the summer of 1884 in the region just beyond that explored by us. A single extract, describing a fiord a short distance north of Ikamiut, furnishes a fitting supplement to the descriptions which we have given.

"The 23d, at noon, we rested on an island in the entrance to a fiord, and the same evening we camped at its farther end. Though the whole length of the fiord was only some twenty-two miles, it surpassed anything we had yet seen in bold mountain scenery.

"It was not unlike the mouth of an immense carnivorous animal, whose teeth were mountains some four to five thousand feet high. As the boat proceeded the scenery changed, and the eye was attracted from one picture to another seeming to surpass it; but it was after having landed on the farthermost shore that the landscape became altogether imposing. The air was

FIG. 25.—The backward view from the glacier down the fiord. Moraines in the foreground.

transparent and calm. In the west the waters merged
into the sky, both resembling an endless space wherein
hills, trees, or islands were reflected as distinctly, both
in outline and colour, as the objects they were a natural
picture of, forming floating masses whose distance from
the eye it was impossible to define, looking near and at
the same time very far.

"A green birch forest was in the shadow on the plain
where our tent was erected, and beyond that mountains
of five thousand feet rose abruptly. Their summits,
golden with the rays of the low sun, contrasted strongly
with the deep blue sky, and, as though to remind us of
the northern latitude of the spot, the ice in their clefts
glittered with a force that emphasized the depth of colour. A mysterious sound floated in the air. It came
from some waterfalls, with clouds of spray flying over
the vigorous greensward at their feet. Never have I
beheld a place coming nearer to the idea which I imagine that our forefathers entertained at Valhalla. Here
was the very eternal day of Valhalla, being unlike the
fleeting one of earth in that the subdued light of midnight heightened the mysteriousness of the place. I was
almost surprised at not meeting the old heroes in person. It would have seemed nothing less than natural
on my wanderings that night, and more than once did
it occur that I suddenly fancied I heard the clash and
clang of swords; but it was only the music of the cas-
cades, and I had to console myself with the persuasion
that the doings of spirits were invisible and inaudible to
my profane senses." *

* A. Riis Carstensen, Two Summers in Greenland, pp. 53, 54.

ICE-SHEET

Length of the Ice-Sheet, 1500 miles; maximum width, 700 miles; area, about 512,000 square miles; altitude of its central tract, 8,000 to 10,000 feet above the sea.

Area of Greenland, about 680,000 square miles, including contiguous small islands, but not the fiords, nor Heilprin Land and Melville Land. Further exploration is needed to show whether Heilprin Land is separated from Greenland by a continuous strait.

100 90 80 70 60 50 40 30 20 10 0 10

GRINNELL LAND
ICE FIELDS
ICE FIELDS
Markham, 1876
Sherard Osborne Fd.
De Long Fiord
Cape Kane
Lockwood and Brainard, 1882
MELVILLE LAND
Independence Bay
Kennedy Ch.
Robeson Ch.
PEARY 1894
PEARY, 1892
Land seen 1775
Land seen 1670
Kane Basin
HUMBOLDT GLACIER
Smith Sound
HAYES, 1860
Ingerhold Gulf
Cape Bismarck
C. Athol
Cape York
MELVILLE BAY
Cape Pansch
Franzi Josef Fiord
PAYER PEAK
PETERMANN'S PEAK
Davy Sound
Hansa Lost
Scoresby Sound
Upernavik
Proven
Umanak Fiord
Nongsoak
Waigat
DISCO ISLAND
Godhavn
DISCO BAY
Jacobshavn
Egedesminde
PEARY, 1886
NORDENSKIÖLD, 1883
1870
N. Ström Fd.
N. Isortok Fd.
Holsteinborg
S. Ström Fd.
Arctic Circle
Sukkertoppen
floe drift of the Hansa's Crew
Cape Hegemann
Cape Dan
Angmagsalik Fiord
Dannebrog's Id.
Godthaab
NANSEN, 1888
Umivik
Illumiut
JENSEN'S NUNATAKS
FREDERIKSHAAB GL.
Frederikshaab
Anoritok
Ivigtut
Julianehaab
Lichtenau
Frederiksdal
Cape Farewell

0 50 100 200 300 400
Scale of Miles.

50 40 30

CHAPTER V.

BEGINNING at the southernmost point of Greenland, about latitude 60°, which corresponds to that of Cape Chidley in Labrador, and to that of the Shetland Islands, Christiania, and St. Petersburg in Europe, Danish Greenland upon the west coast is divided into twelve districts, which we will briefly describe in order : *

1. JULIANSHAAB extends in a west-northwest direction through five degrees of longitude and one of latitude, and has about eighteen hundred square miles of territory uncovered by ice. The most of this, however, consists of inaccessible and barren mountains (some of which are from four to five thousand feet high), capped with local glaciers. Eight or ten fiords extend to the inland ice and receive a few icebergs of small size, while the glacier to the east of the most northern sound, Ikersuak, is set down by Rink among those of the third class, which sends off a considerable number of bergs. In general, however, the edge of the inland ice is at a less distance from the sea in this district than in any other for a thousand miles northward. But this does not prevent the vegetation from being more abundant and the general conditions from being more favour-

---

* In this we can not avoid following largely the descriptions of Rink.

able to civilized life than those found anywhere else in Greenland.

The district of Julianshaab constitutes what was known in the earliest history of Greenland as the Öster-bygd, or Eastern Settlement, which was so long thought by some to be situated on the east coast. Much useless effort was therefore spent in endeavouring to discover an "eastern settlement," which did not exist. But it is now generally acknowledged that the early Norse settlements were here, and that they are now marked by various ruins in the vicinity of the ice front extending through nearly the whole length of the district. Among the most important of the ruins are those along the fiord east of Julianshaab leading to Igaliko, a distance of about twenty-five miles. At Igaliko the ruins are well preserved, and traces of a bridge remain. This is supposed to have been a bishop's farm during the flourishing condition of the Norse settlement. A few miles to the north the border is penetrated by Eric's Fiord, along which remains are numerous, eight settlements having been traced, in one of which there are the ruins of a church. Indeed, this locality seems to have been the most flourishing of all the Norse settlements, and is still capable of supporting a few cattle, there being in all from thirty to forty horned cattle, one hundred goats, and twenty sheep here at the present time, but during the Norse settlements there is the record of a considerable dependence upon cows and sheep for the means of sustenance.

This district contained in 1870 a population of twenty-five hundred and seventy, distributed into between fifty and sixty settlements, the largest of which contained two hundred and twenty-three, but nearly all of them less than one hundred. About one thousand of the natives belong to Moravian communities, whose central station is Lichtenau. This district is reported by Rink to yield about three hundred tons of oil and four thousand seal skins annually, and to furnish one third of the fox skins exported from Greenland. The most southern Moravian settlement is at Frederiksdal, which was established in 1824 for the sake of reaching the few heathen Eskimos from the eastern coast which annually come as far as this for trade. It was here that the crew of the Hansa ended their memorable journey along the east coast in the year 1870.* Frederiksdal is within two hours' walk of the most southern point of Greenland, Cape Farewell being on an island about thirty miles off the coast.

2. THE FREDERIKSHAAB district extends from about latitude 61° to 62° 30', the border running a little west of north. The average width of the land uncovered by ice is about thirty miles, and the highest mountains run up to four thousand feet, these being the ones sighted by us on August 3d. At both the southern and the northern end of this district the interior ice comes down almost to the open sea. In the southern part there are enormous precipices facing the sea, which, with the absence of interior channels, makes it difficult for small boats to pass from the south. According to Rink, there are seven fiords in this district through which the inland ice reaches the water, one of the glaciers being of the second magnitude, and another of the third.

The population numbers about eight hundred, distributed in fifteen settlements, only one of which, Frederikshaab, has more than one hundred. Ivigtut is noted for its cryolite mines, which are chiefly worked by Eu-

---

* See page 62.

FIG. 26.—Front of a Greenland glacier, three miles wide.

ropeans, and the larger part of the products exported to Philadelphia. Between Ivigtut and Frederikshaab travelling by small boats is pre-eminently dangerous, both on account of the unsheltered condition of the shore and of the great number of icebergs which come out of Narsalik Fiord. The Frederikshaab ice blink forms the northern boundary of this district. The inland ice here projects in a tongue several miles wide over a low strip of land almost to the open sea, being separated from it only by its own terminal moraine, which is intersected by numerous streams of water which issue from beneath the glacier, and find their way to the sea by whatever course they can. This also is a difficult place to pass in a small open boat. It was up this portion of the inland ice that Lars Dalager set out upon his expedition toward the east Greenland coast in 1751, but only succeeded in reaching some nunataks about twenty miles from the shore. On the horizon were still other mountains, which he was unable to reach, and which he supposed might be on the eastern shore. It was not until 1878 that this illusion was dispelled. In that year Jensen and Kornerup visited them, and found them to be simply isolated mountain peaks rising from a boundless waste of glacial ice. Many interesting facts concerning these nunataks will appear in a later chapter, giving a more detailed account of the exploration of the inland ice.

But while upon the subject, a few words may here well be added concerning Holst's observations along the border of the Frederikshaab ice blink. Dr. Holst, of Stockholm, in the summer of 1880 skirted the whole coast from Sukkertoppen to Ivigtut in a small boat, making many important observations which had escaped the eyes of other explorers. Among the most significant of these was the discovery of a moraine about twelve miles long and from half a mile to more than a mile in width, extending along the southern side of the great glacier where it approaches the sea. This great mass of morainic material still rests upon ice which has not yet thawed away, but which has melted so unequally that it presents a remarkable series of hills and valleys fifty or more feet high, while the moraine covering does not average more than one or two feet in thickness.*

The ice blink in this region can be seen far out to sea, and presents a more imposing appearance than along any other portion of the coast until one reaches the far north. The slope is here so gradual, and the mountain chain along the border so interrupted that there is nothing to interfere with the vision, so that the eye is permitted to wander over the ice-covered slope to the very limit of the horizon.

3. THE GODTHAAB district extends from latitude 62° 50' to 64° 50'. In the southern part the distance to the inland ice is not great, but gradually increases to about seventy miles in the extreme north. The coast is everywhere bordered by numerous islands, which afford protection to small boats as they navigate the waters. Several large fiords penetrate the borderland to the inland ice. The one known as Baal's River, at whose mouth the town of Godthaab is situated, has a length of seventy-five miles, and branches off into several minor fiords, each of which ends in a glacier of considerable size, and furnishes its quota of small icebergs. The current which characterizes the river, however, is not so much due to the supply of water from the subglacial streams as to the effect of the tide.

---

* See American Naturalist, vol. xxii, p. 707.

To the north of this inlet the land is much lower than upon the south, and affords much good pasturing ground for reindeer. South of the fiord the islands rise to a height of four thousand feet in the near vicinity, but gradually diminish in prominence at greater distances. Nansen, in his famous journey across southern Greenland, came out at the head of Ameralik Fiord, which reaches the ocean a little south of Baal's River. Sadlen Mountain, or the Saddleback, near Godthaab, is one of the most conspicuous landmarks along the coast, being distinctly visible on a clear day from Sukkertoppen, sixty-five or seventy miles away.

The population of the Godthaab district is about one thousand, distributed in fourteen settlements, no one of which has one hundred and fifty inhabitants, though Godthaab is the capital of southern Greenland, where, besides the officials, there are both Danish and Moravian missionaries, and a seminary for the instruction of native catechists and teachers. Here, also, is the residence of the royal inspector and the physician of southern Greenland. Along the inner portion of Baal's River, thirty or forty miles back from the sea, there are numerous ruins of the early Norse settlements; this, in fact, being the Westerbygd, or the Western Settlement of the early historians. Here there were reported to have been ninety farms and two churches; but from the small size of the ruins of the houses—none of them being larger than sixteen by forty feet—it is probable that the inhabitants could only have been a few hundred. Here, also, as in the Eastern Settlement and elsewhere on the western coast, the climate is much milder in the vicinity of the ice border than on the coast itself. The islands along the coast for forty miles north of Godthaab are favourite resorts of the eider duck, and they are so situated as to collect an unusually large amount of drift-wood, upon which the prosperity of the native population so much depends.

4. SUKKERTOPPEN, the next district, extends from latitude 64° 50′ to 66° 10′, presenting throughout its whole length a line of precipitous and lofty headlands, some of them rising to a height of four thousand feet, and many of them being upward of three thousand. Four fiords penetrate the border to the inland ice, namely, Isortok, Ikamiut, Kangerdlugsuatsiak, and Kangerdlugsuak, each extending a distance of from forty to fifty miles back from the sea.*

The population of the district is about eight hundred, distributed in six settlements, of which the largest —Sukkertoppen, containing about four hundred—is also the largest in Greenland. This district is one of the most favoured for the capture of codfish and the collection of eider down. The reindeer pastures, also, were formerly among the most important in Greenland. The South Strömfiord, which forms the northern boundary of the district of Sukkertoppen, is characterized by tidal currents of great violence, rendering it almost impossible for boats to cross except at the turn of the tide. It seems also to be a barrier to the migration of the saddleback seals, which are much more numerous to the south of it than to the north.

5. HOLSTEINBORG extends from latitude 66° 10′ to 67° 40′, and comprises a portion of the outskirts which

---

* Like all Eskimo names, these are significant. Isortok means (as before remarked) "having muddy water": Ikamiut, "an unsheltered bay." Another name for this second inlet is Agpamiut, meaning "place of the auk," on account of the loomeries already described. The other two names are general words for fiords bounded by steep promontories.

FIG. 27.—The scenery looking west from a hill near Sukkertoppen.

is wider between the coast and the inland ice than any other, being about eighty miles upon the average. Three or four fiords extend inland to a considerable distance, but not far enough to reach the ice cap. The surface of the country is much lower than that farther south, and in former times furnished the best of all pastures in Greenland for the reindeer. North Strömfiord, which bounds it upon the north, extends through to the inland ice and, as already remarked, separates south from north Greenland.

The population of the district is about five hundred and fifty, distributed in nine settlements. The village of Holsteinborg is just within the arctic circle, and its harbour is one of the best upon the Greenland coast, there being here a beach upon which ships can run in high tide and undergo repairs when the tide is out. Holsteinborg is the most southern point where the conditions permit the use of sledge dogs, a few of which are found here. The whale fishery, which used to be profitable, is now almost wholly abandoned.

6. EGEDESMINDE is the most southern district of north Greenland, and extends from latitude 67° 40′ to Disco Bay, a distance of about sixty miles. In character the country is much like that of Holsteinborg, but it is more cut up by fiords, which are separated from each other by portages so short that umiaks can easily be transported by inland passages from north to south. Aulatsivik Fiord reaches the inland ice at the point from which Nordenskjöld made both his excursions into the interior.

The population of this district is about one thousand, distributed in twenty-two settlements. In the winter there is good dog-sledging along the edge of Disco Bay, which greatly facilitates communication.

Eider ducks are here specially abundant, and codfish and halibut are caught in large quantities, while the walrus is by no means infrequent.

7. CHRISTIANSHAAB occupies the narrow strip of coast around the southeast angle of Disco Bay, extending as far north as Jakobshavn.

The width of the land in this district is only between twenty and thirty miles, while the highest mountains here are not over fourteen hundred feet, making it the easiest point from which to reach the inland ice. Vegetation is peculiarly luxuriant throughout this district, notwithstanding the proximity both of the coast and the inland ice. But Disco Bay is so narrow that it may almost be reckoned as a fiord, giving to this region the climatic conditions of the inner portion of the " outskirts" farther south.

The population of the district is about five hundred, distributed in eight settlements, of which Claushavn, the largest, has one hundred and twenty-seven. The catch of seal is here obtained almost wholly amid the large icebergs which come out from the Jakobshavn Glacier. The settlement at Christianshaab was made by a son of Hans Egede in 1734. The house built by him is still inhabitable, and is situated at the foot of a hill which is said " in July to be beautifully covered with blue and yellow flowers," while crowberries and blaeberries are abundant even up to a height of over a thousand feet above the sea.

8. JAKOBSHAVN extends from the fiord of that name to the fiord of Torsukatak, latitude 70°, and lies wholly in the rear of Disco Island. The population numbers a little over four hundred, distributed in ten settlements. Seal are especially abundant in Jakobshavn Fiord, through which there is a constant procession of

great icebergs from the magnificent glacier at its head. This is the glacier which was studied so carefully by Helland in the summer of 1875, and whose motion in the central part he found to be about sixty-five feet per day. One iceberg observed by Helland in Jakobshavn Fiord rose three hundred and ninety-six feet above the water. The annual amount of ice carried away in the bergs of this fiord is estimated by Helland to be between 2,900,000,000 and 5,800,000,000 cubic metres. The calving of these bergs is one of the most impressive scenes which it is possible to imagine. We quote a condensed account by Helland :

"Without any previous indication, a tremendous roaring noise was heard, while at the same time a white dust was seen to rise and a large piece of the glacier was detached from its outer edge, which, after having rolled for some moments in the water, reared its edge in the air, but almost instantly the pinnacle top of this edge burst asunder and crumbled while falling. The calving having thus commenced, it was instantly followed by a much larger piece being detached and issuing from the middle part of the glacier at the rate of one metre per second. But the extensive *bouleversements* which now ensued made it impossible to discern the number and size of the larger bergs which were formed out of this portion of the glacier, because clouds of dust now arose in different places, and the floating bergs in the vicinity were also put in motion, rolling and calving. It was more than half an hour before the whole scene again was calm and the thundering noise which had accompanied the disturbances had subsided."*

---

* See Rink's Danish Greenland, p. 364.

9. GODHAVN comprises the southern and western shores of Disco Island, and its principal trading station, officially known also as Godhavn, but called by the English sailors Lievely, is the point most frequently visited by explorers and whalers on their way farther north. The island of Disco contains an area of about two thousand square miles, the most of which is a tableland from two thousand to four thousand feet above the sea, and enveloped in perpetual snow and ice. This area is largely covered with beds of basalt and sandstone, greatly in contrast with the rocks of southern Greenland, which are wholly granitic in their character. Extensive beds of coal also occur.

The population numbers about two hundred and fifty, distributed in seven settlements. There are but few harbours on the island, and in winter the bay is not frozen over with sufficient permanence to render sledging at all safe, making the isolation extreme during that portion of the year.

10. RITENBENK occupies both sides of Waigat Strait, which separates Disco Island from Nugsuak Peninsula on the north. The land upon this peninsula rises to a height of seven thousand feet, and the Waigat is bordered upon both sides by precipitous walls from three thousand to four thousand feet high. The population numbers about four hundred and fifty, distributed in eight settlements. At the extreme end of Nugsuak Peninsula there is situated the only ancient Norse ruin in north Greenland. It is of stone, about ten feet long and six feet broad, and is built with a skill far greater than any ever attained by the Eskimos.

The highest mountains of western Greenland are found in the vicinity of Disco and Umanak, and the rocks are of much more recent geological formation

9

than in the southern part, including extensive areas of
sandstone, shale, and basalt of Cretaceous and Tertiary
age. Over this area the mountain masses, reaching a
height of five thousand feet, are tabular in form and
rise to their summits by a series of terraces. Granitic
rocks in some instances occur at the base, but in the
sandstone strata the edges of extensive coal beds appear.
In these, many trunks of tall trees, according to Rink,
" are still standing upright, with the remains of their
roots inserted in the very soil that gave growth to them.
. . . Perfect and complete impressions of leaves have
been discovered in abundance in the surrounding rocks.
Fruits, seeds, and even remains of insects are among
these striking proofs of an ancient flora and fauna
which to all appearances must have required a climate
like that of middle Europe at the present day. The
coal beds also afford a striking evidence of the igneous
origin of the superincumbent trap which on bursting
out made its way through the said deposits. The coal
is distinctly seen to have been altered in various degrees by the heat from the melted masses. In one
instance a small trap eruption crosses and spreads over
a thin coal bed for some extent. The coal in immediate contact with the volcanic rock was found to be totally deprived of its volatile bituminous ingredients and
changed into coke. In another place a coal bed was
found converted into anthracite ; and, lastly, a most remarkable bed of graphite has been discovered which
leaves no doubt of its having originated in the same
way, the heating and metamorphosing action having
here reached a higher degree of intensity." *

The basalt beds in this region, occupying an area of

---

* Danish Greenland, pp. 77, 78.

about five thousand square miles, are mostly horizontal and from fifty to one hundred feet in thickness, and are separated from each other by sheets of reddish clay and tufa containing angular fragments of lava. It is this alternation of structure which occasions the terraced appearance of the mountain sides of the region. According to Rink also, these volcanic rocks " show a series of varieties gradually passing from the granular or crystalline to a more compact basaltic texture, with a tendency at the same time to become vesicular or even spongy, so as to assume the appearance of true lava." *

Geologically the coal beds of Greenland are much later than the Carboniferous period. The accompanying plants indicate that some of them belong to the Upper Cretaceous and others to the Middle Tertiary (Miocene). " Of the first, two subdivisions have been observed, of which the lower contained fifty-six species of plants, among which nineteen are ferns and nine cycadeæ. The Tertiary group is most beautifully displayed at Atanekerdluk (mainland, 70° north latitude), the number of species determined amounting to one hundred and sixty nine. The sandstone beds with their subordinate beds have a thickness of several hundred, rising even to two thousand, but not exceeding twenty-five hundred feet. In following them along the shores they are not always found to contain coal beds, but always layers of claystone and slate. Large pieces of fossil wood are scattered over the Asakak Glacier near Umanak." †

The Tertiary beds in this region bear striking witness to the changes of climate which the region has experienced, and to the fact that there is a lineal connec-

---

* Danish Greenland. p. 386.   † Ibid., pp. 384, 385.

tion between the present flora of the north temperate
zone and the ancient arctic flora of Greenland. During
the middle portion of the Tertiary period the climate of
north Greenland corresponded closely with that which
now exists in Virginia and North Carolina. As enumerated by Asa Gray, the familiar plants found in these
beds comprise " magnolias, sassafras, hickories, gum
trees, our identical southern cypress (for all we can see
of difference), and especially sequoias—not only the two
which obviously answer to the two big trees now peculiar to California, but several others; they equally comprise trees now peculiar to Japan and China—three
kinds of gingko trees, for instance, one of them not
evidently distinguishable from the Japan species which
alone survives. We have evidence not merely of pines
and maples, birches, lindens, and whatever characterize
the temperate-zone forests of our era, but also of particular species of these so like those of our own time and
country that we may fairly reckon them as ancestors of
several of ours." *

11. UMANAK is the district occupying the peninsulas and islands surrounding the broad fiord or bay of
that name, and which extends from about 70° 45' to 71°
45'. Into this bay two glaciers of the first magnitude
pour their vast volumes of ice, which is increased by
that of several others of smaller magnitude. The general character of Nugsuak Peninsula upon the south is
like that of Disco Island, basalt and sandstone abounding on the high table-lands. The bay of Umanak affords
the best opportunities for seal hunting anywhere along
the coast. The number of inhabitants is about eight
hundred, distributed in fifteen settlements. In winter

---

* Scientific Papers of Asa Gray, vol. ii, p. 227.

the water is frozen over, so that travel by dog sledges is safe, rendering that season of the year by no means undesirable.

Umanak Fiord furnishes a good proportion of the icebergs which find their way south with the Labrador current.

12. UPERNIVIK extends from Umanak Fiord to Tasiusak, latitude 73° 24', the most northern trading point in Danish Greenland. The peninsula Swarte-Huk, comprising the larger part of the territory, is but little known. In the vicinity of the settlement of Upernivik the inland ice comes down to the water in a magnificent glacier, which was found by the Swedish surveyors to be moving in the month of August at the rate of a hundred feet per day. In winter the sun is seventy-nine days below the horizon in this latitude, and there is no communication with the outside world except by a single ship in the summer and a single sledge post which comes up from Umanak Fiord in winter.

From Upernivik to Cape York in a direct line is about two hundred and fifty miles, but following the shores of Melville Bay, forming the two other sides of the triangle, the distance is nearly twice as great. This part of Greenland is almost as little explored and as difficult of approach as that in the extreme north. Everywhere the ice seems to come down to the water's edge, and the ice pack in the bay is so irregular in movement that it is the great terror of navigators. The Devil's Thumb is a peculiar landmark in the vicinity of Tasiusak, two thousand or three thousand feet high, which is usually the last land seen until the promontory of Cape York comes into view, in about latitude 76°. So far as known, the shores of Melville Bay are uninhabitable both to men and to animals.

But from Cape York northward to Littleton Island, in Smith Sound, numerous bays indent the coast, and areas of considerable extent are free from snow in summer, so that a few musk oxen and reindeer can maintain existence. Accompanying these, and depending largely upon them, are a few hundred heathen Eskimos, about whom little was known until Kane's contact with them in the winter of 1853, when they befriended him and did much to save his party from destruction. Subsequent expeditions have confirmed the good impression which Kane received.

Inglefield Gulf, setting back into the interior from Whale Sound, between latitude 77° and 78°, has been the centre from which Peary has set out upon his expeditions for the survey of the northern coast of Greenland. In general this gulf is bordered by promontories one thousand or two thousand feet in height, from whose base the snow melts off in summer; but at seven or eight points the coast line is interrupted by fiordlike depressions through which projections from the inland ice reach the water level and send off a limited number of small icebergs. The most of these glaciers, however, are very sluggish in their movements as compared with those in southern Greenland.

Humboldt Glacier, discovered by Kane, enters Kane Basin beyond Smith Sound, its perpendicular front occupying nearly the whole space between latitude 79° and 80°. In Peary's first expedition from Inglefield Gulf he encountered considerable difficulty from the fact that the route chosen was so near the Humboldt Glacier that his progress was obstructed by the crevasses occasioned by the movement of ice toward it. To escape the difficulty he went farther into the interior, where he had no more trouble.

Large glaciers come into Kennedy Channel through Petermann's Fiord, in latitude 81°, and others still larger into Sherard Osborne Fiord, in latitude 82°.

In 1882 Lockwood and Brainard, of Greely's party, reached the highest point yet attained by any one, in latitude 83° 24', but at a season of the year when little could be determined concerning the nature of the country. Lieutenant Peary in 1892, however, after making a journey of five hundred miles across the inland ice in midsummer, came down upon the northeastern coast of Greenland, in latitude 82°, longitude 40° west from Greenwich, and surveyed the coast for a distance of seventy or eighty miles. He found conditions very similar to those in southern Greenland, namely, a line of promontories, rising in some cases to thirty-five hundred feet, between which there were numerous depressions down which great glaciers were moving northward into open water. This he called Independence Bay, which was bordered upon the north by a long stretch of high land free from snow upon the 4th of July. The Academy Glacier, moving northward into this bay, is several miles in width, and debouches into the water with all the characteristics of the glaciers described in Disco Bay and Umanak Fiord. On this northern border of Greenland, in midsummer, Peary was greeted by the sight of green grass and beautiful flowers and butterflies, and of herds of musk oxen revelling in the luxuriance; and while lying down upon this carpet of beauty to rest, in the middle of the day, his ears were greeted with the familiar buzz of the bumblebee and the less pleasant hum of swarming mosquitoes, and this within five hundred miles of the north pole.

As already said, the east coast of Greenland is much more inaccessible and much less known than the west coast. In 1822 Scoresby touched upon the east coast between latitude 74° and 75°, but it was not until 1870 that any considerable extent of that part of it was explored. During that year Koldewey reached Cape Bismarck, about latitude 77°, and continued his course as far down as Cape Brewster, latitude 70°. No Eskimos were found along this portion of the coast, but numerous remains of deserted igloos occurred in various places, and large herds of musk oxen and reindeer were seen. Franz Josef Fiord penetrates the interior for nearly one hundred miles, and twenty or thirty miles farther westward Petermann's Peak rises from the icy waste to a height of about eleven thousand feet, being the highest mountain yet discovered in Greenland.

From latitude 70° to 65° scarcely anything is known of the condition of the border. The Spitzbergen ice pack is here so dense along the Greenland coast that access to the shore has never been obtained. Slight glimpses of it were obtained by the crew of the Hansa as they drifted past it upon the ice in the winter of 1870, but nobody has succeeded in actually exploring it.

From Cape Dan to Cape Farewell the coast is better known, having been explored in 1828 by Captain W. A. Graah, of the Danish Royal Navy, who was despatched thither to search for relics of the lost colonies in the East Bygd, which was then supposed to have been located upon that part of the coast.

Captain Graah, setting out from Frederiksdal early in the spring, with native help succeeded in reaching Dannebrog's Island (latitude 65° 15') late in August, only a short distance south of Cape Dan; but the season was so far advanced that he was compelled to spend the winter upon the coast, which he did at Nukarvik (latitude 63° 15'). Little has since been learned in addition to Graah's account, though in 1884 Captain Holm reached Angmagsalik, a short distance north of Cape Dan, and spent the winter there with a colony of heathen Eskimos. Still later, Nansen, when released from the floe ice, went over a portion of the same territory to find a suitable starting point for his famous journey across southern Greenland in 1888. From the description of these explorers it appears that the eastern coast has a much narrower belt free from ice than the western has, and is capable of supporting a much smaller population. There are now about five hundred Eskimos along this coast, and it would not seem possible for the number to be materially increased. All these explorers report an absence of reindeer from the region, and, indeed, a total ignorance of the animal on the part of the natives. This portion of the coast is beset with floe ice the whole year round, but does not seem to be much more encumbered with glaciers reaching the sea than is the west coast, the most formidable of these being Puisortok, about latitude 62°—a glacier presenting about the same appearance to Graah as it did to Nansen sixty years later.

Graah's description of the coast about the sound of Ekallumiut, where he wintered in latitude 63° 30', is worthy of reproduction.

"August 30. The place we now were at was the Ekallumiut so often mentioned. The cove, the length of which is between one and two cable lengths, has on both sides of it, but particularly on the eastern, fields of considerable extent covered with dwarf willows, juniperberry, black crakeberry, and whortleberry heath, the first-named growing to the height of two feet, and the whole interspersed with a good many patches of a fine

species of grass, which, however, was very much burned
by the heat of the sun, except in the immediate vicinity
of the brooks and rivulets that in great number ran
down the sides of the hill and intersected the level land
in every direction. At the bottom of the cove stretches
an extensive valley, through which runs a stream
abounding in char [salmon] and having its source in
the glaciers, of which several gigantic arms reach down
into the valley from the height in the background. On
the banks of this brook the grass grew luxuriantly, but
it was far from being, in many places, of a height fit for
mowing; so that even this spot, where grass was more
abundant than anywhere else, perhaps, along the whole
coast, does not seem calculated to furnish winter fodder
for any considerable number of cattle. Various flowers,
among which was the sweet-smelling lychnis, everywhere
adorned the fields. The cliffs recede to a distance of
from two hundred to three hundred paces from the sea,
rising then, however, almost perpendicularly far beyond
the ordinary height, the clouds seeming to rest upon
their snow-clad summits. Rock and ice slides are here
events of frequent occurrence. Down a ravine on the
southwest side of the cove, particularly, huge masses of
ice were every moment precipitated, crumbling in their
fall to dust and accompanied with a noise like thunder.
At this really beautiful spot the natives of the country
round assemble for a few days during their brief summer to feast upon the char, that are to be got here in
great plenty and of great size, the black crakeberry and
angelica, and to lay in a stock of them for winter use,
and give themselves up to mirth and merrymaking.
This evening they collected together in a body of some
two hundred or two hundred and fifty persons and began by torchlight their tamboureen dance, a festivity to

which I was invited by frequent messages sent me during the night, but in which I was prevented by a slight attack of fever from taking part." *

Of the general climatic conditions in southwestern Greenland Dr. Henry Rink has given us a very comprehensive account. According to him, a short distance, in progressing from the extreme border to the inland ice, produces a great climatic change. In almost every day of summer, when the weather is warm and sunny along the outer shore, the temperature will suddenly fall to within two or three degrees of the freezing point. But this change does not usually extend farther into the interior, bright sunshine continuing to characterize a considerable belt of country bordering the ice cap. Dense clouds encompass the headlands only to be dissolved within two or three miles of the border. To pass into the interior at this time of year is to change from winter to summer. In the winter, however, conditions are reversed, the temperature being much lower in the interior portions of the border than along its outer edge, the difference being more than ten degrees in favour of the oceanic border.

The mean temperature at Julianshaab, near Cape Farewell, is 33° F., while that at Upernivik is 13°, a difference of twenty degrees ; but the average summer temperature at Upernivik is only ten degrees below that of Julianshaab, the former being 38° and the latter 48°. In winter, however, there is an average difference of twenty-seven degrees, namely, 7° at Upernivik and 20° at Julianshaab. Thus it appears that the climatic conditions in southern Greenland are remarkably uniform, the

* Graah's Narrative of an Expedition to the East Coast of Greenland. pp. 106. 107.

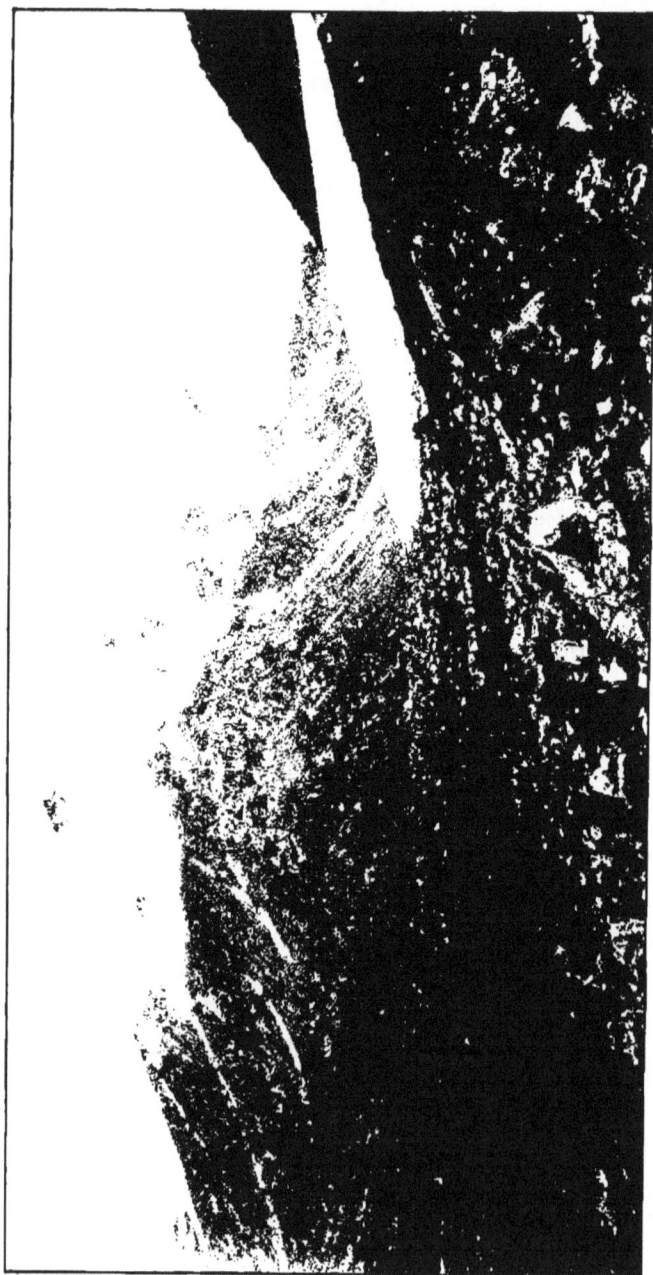

Fig. 28.—A side view in the outskirts of Greenland.

winters being less severe than one would have supposed, while the summer is truly arctic. This is doubtless caused by the equalizing effect of the large areas of water surrounding the southern portion. It should be said, however, that the observations at Julianshaab have all been made upon the extreme border, and we are ignorant of the climatic conditions in the interior of the southern part. But in latitude 64°, at Umanak, in the interior of the Godthaab Fiord, while the mean annual temperature is nearly the same as that at Godthaab, on the border, the summer is much warmer and the winter much colder than at Godthaab, the mean temperature of July being four degrees higher, and that of December six and a half lower. Still the vegetation, even in the interior, always betrays the arctic character of the climate.

The extremes of heat and cold, which have so decided an effect upon vegetation, are also discussed by Rink with much fulness of detail. At Lichtenau (latitude 60° 31'), during four years of observation, the thermometer only once rose as high as 66°, and only four times above 60°; while at Upernivik (latitude 72° 48') the thermometer sometimes rises to 59°. At Julianshaab (latitude 60° 43') the thermometer rose to 68° in the summers of both 1853 and 1854, while in Disco Bay (latitude 68° 48') the thermometer rose to 64° on June 28, 1858.

Like the central part of Switzerland, which has its warm, dry, *foehn* wind, sometimes called the "snow eater," descending from the upper Alps in the autumn and winter, and like the northwestern part of the United States, which has its chinook wind pouring from the west into the interior over the Cascade and Rocky Mountains, and suddenly melting off the snows and drying up

the land surfaces, Greenland has its mysterious warm
currents flowing down from the interior to the outskirts,
and producing sudden and most remarkable changes in
the temperature. Frequently these will bring on an extensive thaw in January and February. Nor are they
limited to the southern latitudes. In the most northern
settlements these winds have been known in the midst
of winter to raise the temperature for a short time as
high as 42°. As an effect of these winds, we have such
paradoxes as the following : " During eight consecutive
days of the long arctic night in November and December, 1875, it was warmer at Jakobshavn (latitude 69°
20'), in western Greenland, than in northern Italy ; and
for part of this time Upernivik, though in continuous
winter darkness, was warmer than the south of France." *

The cause of such winds has been assigned by meteorologists to the effect of the latent heat set free by the
condensation of the moisture as the winds have passed
over high mountains, or, as in the case of Greenland,
over large and high snow-covered areas, with an ensuing
rapid descent to lowlands. Greely records that during
such a wind the barometer rose a quarter of an inch during a single day. The direction of these warm land winds
varies according to locality, but they always come down
from the ice cap, and are felt most at the head of the fiords.
Sometimes they come down with such sudden force between the walls over the fiords as to sweep everything
before them, and to raise columns of fog, while at other
times they come as a gentle breeze with a clear sky.
Occasionally they are accompanied by heavy rains, but
usually, as in Switzerland, when blowing for several
days they are extremely dry, so that they evaporate

---

* Nature, vol. xvi, pp. 294, 295. (August 9, 1877.)

nearly all the moisture produced by their melting power.

The extremes of the winter cold in Danish Greenland are given by Rink as —29° F. at Julianshaab, in 1863, and —42° at Umanak, and —47° in Upernivik. At Upernivik, also, the temperature sometimes falls below the freezing point in July, once reaching the extreme of 27¼°, but in south Greenland the temperature rarely reaches the freezing point in July and August. Among the instances of extreme changeableness of weather mentioned by Rink are the following :

"On December 26, 1819, in 64° north latitude, a heavy rain was pouring down incessantly the whole day, with a calm and steady thermometer from 54° to 57°, the mean of that month amounting to 26½° in the same year. On May 22, 1850, the author found a saxifrage blossoming very beautifully in 70° 30' north latitude. On the last day of June and the first of July, 1854, after a severe winter, he visited the headland of Nunar-suit, about six hundred miles farther south, and traversed it on foot. The smaller inlets, as well as the lakes slightly above the level of the sea, were not only covered with ice, but the latter, as well as the adjoining country, was covered with snow to such a degree that the border between the ice and land was levelled and quite imperceptible. At Julianshaab, in February, 1855, the warm land wind set in with light breezes and a temperature of 32°, clearing the sky and lasting for about a fortnight, with beautiful weather, but was in March and April succeeded by heavy snowfalls, and on May 1st the gardens were still covered with a sheet of snow five feet thick. In the first week of June it snowed continuously for thirty-six hours, so as to make the roads between the houses almost impassable. Nevertheless, the summer

turned out unusually fine, and in all respects favourable
to vegetation. In 1863 and 1864 the winters in regard to
severity surpassed any of which the author has ever been
able to acquire information from earlier accounts, and
possibly we should have to go back a whole century to
find their equals. In 64° north latitude not a single
drop of rain was noticed from September 27, 1862, to
the 20th of May next, on which day the snow had obtained a height of from eight to twenty feet between
the houses. At the southernmost settlement, during
six days in March, the thermometer did not rise above
—17½°. The succeeding winter was almost identical as
regards the whole amount of cold, but the period of extreme cold both commenced and abated somewhat earlier. As regards the quantity of snow, a more recent
winter has nevertheless surpassed those here mentioned."*

* Danish Greenland, pp. 62, 63.

CHAPTER VI.

In their distribution and history, no less than in their habits, the Eskimos are a most singular people. A small colony of them lives in northeastern Asia, west of Behring Strait, but the largest proportion, about twenty thousand, is found in northwestern Alaska. From there they extend in inconsiderable numbers eastward along the northern coast of British America, to Baffin Bay, and down the coast of Labrador to the Strait of Belle Isle. Western Greenland, however, affords support to the great bulk of the race in eastern America, about ten thousand being at the present time found there.

The word *Eskimo* means "meat-eater," and was given to the race to describe their habitual diet, which is determined for them by the necessities of their situation; for even in southern Greenland there is scarcely any vegetable food attainable. A few berries and the stalks of angelica, a plant which somewhat resembles celery, furnish the only relief which the people have from a pure diet of flesh; though, when driven by hard necessity, they sometimes subsist for a while upon a species of seaweed, which is abundant. North of Melville Bay, however, the people live entirely upon meat, the most of which is eaten raw. The little which is cooked is so prepared for the sake of extracting the blood, which, with water, forms their only drink. Thus

10   127

it would seem that the ordinary appellation of the people is so descriptive that it ought to be satisfactory; but it is far from being so, and is very much disliked by natives. They prefer to be called, and call themselves, *Innuit*—that is, "The People." In their own estimation, they are the people and we the barbarians.

Their origin and development are still as great mysteries as ever. Linguistically they belong to the agglutinative family, in which the prefix, the stem, and the suffix of the words retain their individuality, and refuse perfectly to coalesce, as they do in the inflected languages. To this extent the Eskimos are affiliated with the North American Indians, and at the same time to the Turanian nations, of which the Hungarians, the Finns, the Turks, and the Tamils are examples. A marked tendency in these languages is the compression into one word of what would require a whole sentence in other tongues. For example, seventeen words are used in English to express the idea in the following sentence: "He says that you also will go away quickly in like manner and buy a pretty knife." But in Eskimo this is all expressed in a single compound word: Savigiksiniariartokasuaromaryotittogog. Analyzed, this is: *Savig*, a knife; *ik*, pretty; *sini*, buy; *ariartok*, go away; *asuar*, hasten; *omar*, wilt; *y*, in like manner; *otit*, thou; *tog*, also; *og*, he says.* Similarly, when the Ojibway Indian wishes to refer to "those persons who came this way and did him and me a favour," he needs but the one compound word, Gahpemeezheshahwaynemeyungidechig.†

---

* Robert Brown, in Encyclopædia Britannica.
† Rev. Sela G. Wright, a missionary among the Ojibway Indians.

But beyond such general resemblance there is little similarity between their language and that of the American Indians, there being no common root words. It would be as difficult for an Eskimo to understand an Indian as for a Chinaman to understand an Arab. On the other hand, the uniformity of the Eskimo dialects is at once surprising and significant. Though separated by thousands of miles of inhospitable arctic wastes, which must have rendered intercommunication well-nigh impossible for many centuries, the Eskimos of Greenland, Labrador, Alaska, and eastern Siberia speak the same language with less dialectic variation than can be found in different counties of England.

The investigations of Lewis H. Morgan * concerning the modes of reckoning relationship, involving fundamental conceptions of social life, brought out clearly the fact, otherwise rendered probable, that the Indian and the Eskimo must have branched off from the racial stem in early prehistoric times. The Eskimo does not, like the Indian, consider his cousins as his brothers, nor call his nephews and nieces sons and daughters; but he does call the husbands and wives of nieces and nephews sons-in-law and daughters-in-law, and the husbands and wives of cousins are regarded as brothers-in-law and sisters-in-law. The Eskimo conception of the family and the tribe seems thus to be a connecting link between that of the Aryan and that of the Turanian race. The antiquity of this conception is no doubt very great.

The primitive religion of the Eskimos does not differ essentially from that of the American Indian. With

---

* Systems of Consanguinity and Affinity of the Human Family, in Smithsonian Contributions to Knowledge, Washington, 1870, No. 218, pp. 267–277.

both there is belief in the immortality of the soul and of a Supreme Being; while with both the chief means of influencing the Supreme Being is the intercession of the *angekok*, as he is called by the Eskimos, or the "medicine man," as he is denominated among the Indians. Originally their whole social organization was closely dependent upon the influence of these religious leaders. Nor has the substitution of Christianity for the original religion in Greenland wholly effaced the primitive habits of religious thought.

The Supreme Being of the Eskimos was known as *Tornarsuk*, under whom there were numerous subordinate *tornaks*, or guardian spirits, through whom only supplications for aid were lawful. There was, however, another method of invoking supernatural aid, which was unlawful, though effective, and which corresponded to witchcraft, so common among all nations. Although their main dependence was upon the intercession of the *angekoks*, men were supposed to be aided in securing assistance from *Tornarsuk* by the use of prayers and amulets. Health and long life, for example, were thought to be conferred upon a child by preserving his navel string for occasional use as an amulet. Pieces of old wood, stones, bones, bills and claws of birds, and other things also served as amulets. Many articles of commerce are used by them secretly for the same purpose. Probably, however, there is no greater superstition in an Eskimo who carries around a kernel of coffee in his pocket to secure long life than there is in the Anglo-Saxon who buries a bean under a stump for the purpose of removing warts from his hands.

According to Egede, the "science of the *angekoks*" consisted of three things: First, "that he mutters certain spells over sick people, in order to make them

recover their former health "; second, " that he com-
munes with *Tornarsuk*, and from him receives instructions to advise people what course they are to take in
affairs, that they may have success and prosper therein ";
third, " that he is by the same informed of the time and
cause of anybody's death, or for what reason anybody
comes to an untimely and uncommon end, and if any
fatality shall befall a man." *

It is difficult to tell how much of the influence of
the *angekoks* was due to rank imposture, and how much
to real mental and moral superiority. On the one hand,
the early missionaries believed the *angekoks* to be im-
postors pure and simple. According to Egede, the
angekoks supported their claim to the power of visiting
the unseen world by transparent tricks. Assembling
the spectators in one of the houses after dark, when
every one is seated the "*angekok* causes himself to be
tied, his head between his legs and his hands behind
his back, and a drum is laid at his side; thereupon,
after the windows are shut and the light is put out, the
assembly sings a ditty, which, they say, is the composition of their ancestors. When they have done singing
the *angekok* begins with conjuring, muttering, and
brawling; invokes *Tornarsuk*, who instantly presents
himself, and converses with him. . . . In the meanwhile he works himself loose, and, as they believe,
mounts up into heaven through the roof of the house,
and passes through the air till he arrives into the highest of the heavens, where the souls of " the chief of the
angekoks reside, by whom he gets information of all he
wants to know. In the course of his career the successful *angekok* passes through various other analogous ex-

---

* Hans Egede's Description of Greenland (London, 1818), p. 188.

periences before he arrives at a point of supreme influence among his fellows. This, as we have said, is set down by Egede as pure imposture.

But it may well be maintained that these experiences are often genuine delusions, such as occur in hypnotic experiments of modern times. At any rate, it is the opinion of many who have been intimate with the Eskimos that the *angekoks* are usually persons of more than ordinary intelligence and respectability, and hold their position by reason of real personal merit. Dalager describes one who lived near him as a person who seemed sincerely to seek information about God and his work, and to lead, on the whole, an exemplary life; and Rink also grants that the misuse of authority to promote selfish aims by the *angekoks* exhibits nothing which peculiarly distinguishes them from the priestly class in other nations.

In the Christian transformation which has taken place in Greenland, the Supreme Being of the Eskimos has been changed into his Satanic Majesty, and all the work of the *angekok* has been degraded, in the estimation of the people, to the level of witchcraft. This leaves no room in later times for the original conception of the devil, which was that of an old woman by the name of *Arnarkuagsak*, who " resided in the depths of the ocean, ruling over all the inhabitants of the sea, and was made the grandmother of the devil."

In their conceptions of the spiritual world, it was divided into two abodes—an under world and an upper world; but it is not altogether strange that their pictures of these abodes differ radically from those drawn by the inhabitants of more sunny climes. According to the Eskimos, the under world is much the pleasanter, being a delicious country, " where the sun shines continually with an inexhaustible stock of all sorts of choice provisions"; whereas "cold and hunger are encountered in the upper world." Still, neither of the worlds is thought to be wholly unendurable.

The Eskimos, too, had their legends concerning the origin of things. Of these we have room for only two or three specimens. At the beginning, a Greenlander sprang out of the ground, and dug a wife out of a little hillock. From this couple the Eskimos were descended. The origin of foreigners was less noble. A Greenland woman gave birth to both children and whelps; "these last she put into an old shoe and committed them to the mercy of the waves, with these words, 'Get ye gone from hence, and grow up to be Kablunaets.' This, they say, is the reason why the Kablunaets always live upon the sea; and the ships, they say, have the very same shape as their shoes, being round before and behind." *

The sun, moon, and stars were regarded as ancestors of the Eskimos, who, for various reasons, had been lifted up to heaven. The moon was thought to be a young man who had fallen deeply in love with his sister, the sun, and who "was used every night to put out the light that he might caress her undiscovered; but she, not liking these stolen caresses, once blackened her hands with soot, that she might mark the hands, face, and clothes of her unknown lover who in the dark made addresses to her, and by that discover who he was; hence, they say, come the spots that are observed in the moon; for, as he wore a coat of a fine white reindeer skin, it was all over besmeared with soot; whereupon Malina, or the sun, went out to light a bit of moss; Anningate, or the moon, did the same, but the flame of

* Description of Greenland, p. 199.

his moss was extinguished; this makes the moon look like a fiery coal, and not shine so bright as the sun. The moon then ran after the sun round about the house to catch her; but she, to get rid of him, flew up into the air, and the moon, pursuing her, did likewise; and thus they still continue to pursue one another, though the sun's career is much above that of the moon." *

Physically the Eskimos have considerable resemblance to the Chinese. Nansen describes † them as having round, broad faces, with small dark eyes, sometimes slightly resembling those of the Mongolian races, a flat nose, rather narrow between the eyes, a broad mouth with heavy jaws, and as being in general short of stature and, owing to their mode of life, deficient in the development of their lower limbs. They wash so seldom that it is difficult to determine the natural colour of their skin, which is really a brownish or a grayish yellow, though much intensified by the accumulation of dirt.

An incident related to me by Dr. Dudley P. Allen is significant in its bearing upon the question of the general race affiliation of the Eskimos. Dr. Allen is Professor of Surgery in the medical department of the Western Reserve University, Cleveland, Ohio, and has charge of one of the hospitals of the city. Among the patients which came under his care last summer was an Eskimo from Labrador, who had been at the Columbian Exposition in Chicago. While passing through Cleveland on his way home he received an injury and was detained for some time in the hospital. Among Dr. Allen's guests during the summer was Dr. Kerr, of Canton, China, who for forty years had been chief of the hospital estab-

* Description of Greenland, pp. 207, 208.
† The First Crossing of Greenland, pp. 172, 173.

lished there by Presbyterian missionaries, and hence by his long residence and practice had become perfectly

FIG. 29.—A typical Eskimo couple.

familiar with Chinese physiognomy. As these two were passing through the wards of the hospital in Cleveland, the face of the Eskimo attracted the attention of Dr. Kerr, and his instantaneous remark was, " Why, here is a Chinaman! What's the matter with him?" Such an impression made on one so familiar with the Chinese speaks more strongly of the Asiatic affiliations of the Eskimos than a volume of statistics would do. It would be difficult to believe that any one would ever mistake an Indian for a Chinaman, but the resemblance between the Eskimos and the Chinese, while largely indefinable, is as clear as that which impresses all tourists between the tribes of southeastern Alaska and the Japanese.

Dr. F. A. Cook, ethnologist of the first Peary North Greenland Expedition, has made the most careful study of the Eskimos north of that region. From his measurements we learn that " the average male is five feet one inch and a half in height, and his average weight is one hundred and thirty-five pounds, while the woman is four feet eight inches in height and has an average weight of one hundred and eighteen pounds." Dr. Cook remarks also upon the Mongolian cast of their countenance, and that " the muscular outlines of the body are nearly obliterated from the fact that they have immediately beneath the skin a layer of blubber or areolar tissue which protects them against extreme cold."

It is not strange that the Eskimos have been generally regarded as an Asiatic race which sought to better its condition by emigrating to America at a somewhat recent period; * for, in addition to the resemblance, both physical and linguistic, between the Eskimos and

---

* See especially Mr. C. R. Markham's paper, printed in the Arctic Papers prepared by the Geographical Society of London in 1875 for the Nares Expedition.

the various Asiatic tribes, this view is supported by
the surprising uniformity already referred to in the
language of the Eskimos, however widely scattered;
by a similar uniformity in their arts, traditions, and

FIG. 30.—A company of Eskimo boys.

folk-lore; and by the fact that a few Eskimos still
live in Siberia. It is cogently argued that this uniformity could not have been maintained if the Greenland Eskimos had been separated from the Alaskan
stock for many centuries. The late immigration of
the Greenland Eskimos is thought to be confirmed
also by the historical evidence that the Europeans,
who settled in southern Greenland about the year
1000, did not come in contact with any natives for
two or three hundred years, and then only in northern
Greenland.

On the other hand, Dr. Rink * maintains (and he is supported by the high authority of William H. Dall) † that the Eskimos are an aboriginal American race which, yielding to the pressure of the more favoured central tribes, was pushed northward down the principal water courses of British America to the arctic border of the continent, where the conditions of life necessitated the abrupt change in the habits of the people which now separate them from all other races in the world. This theory naturally accounts for the enmity which exists between the Eskimo and the Indian whenever they come in contact with each other; but it does not so well account for the clear line of demarcation between them in their general appearance, their systems of social life, and their linguistic characteristics.

On the whole, therefore, it would seem that the Eskimo emigration from Asia was independent of the Indian, but that it occurred at a much more remote period than Mr. Markham supposes. Still the facilities for migration have always existed in the vicinity of Behring Strait; for so narrow is this channel that boat loads of the natives living upon its shores now annually cross and recross from one side to the other, while the water is so shallow that an elevation of only one hundred and seventy-one feet would establish land connection between the continents at the present time. Such connection was probably in existence at no very remote period.‡

---

* Meddeleser om Grönland, Ellevete Hefte, 1887. Also Danish Greenland. pp. 404, 405.

† See Contributions to North American Ethnology, United States Geographical and Geological Survey of the Rocky Mountains Region, 1877, vol. i, pp. 93–106.

‡ W. H. Dall, American Journal of Science, February, 1881.

The Eskimos never live far away from the water, and, with one exception, the tribes are all dependent on their skill in the construction and use of boats peculiarly fitted to ride the breakers of the sea. That exception is the Highland Eskimo of northern Greenland, where the reign of frost is so supreme that boats are almost wholly discarded, and hunting and fishing are prosecuted mainly upon the ice and upon the narrow margin of land, which continues to support small colonies of musk ox and reindeer.

FIG. 31.—A umiak, with women rowing.

The boats of the Eskimos are most ingeniously constructed, being covered with skins of the seal, stretched over a light, strong framework of wood. These are of two patterns, the umiak and the kayak.

The umiak, or women's boat, requires from fifteen to twenty skins for its construction, and is from twenty-five to thirty-seven feet long, about five feet broad and two and a half feet deep, flat-bottomed, and open at the top. The largest of them can carry more than three tons, or six thousand pounds of freight. So light are they that six or eight men can carry one with ease on their backs. The umiak is universally rowed by women, and is chiefly used to transport the families in summer from one hunting station to another, great care being taken not to expose it to rough seas, which it is not adapted to encounter. When thoroughly wet the skins of the umiak become transparent, so that the water can be seen through the bottom and sides as the boat rushes through it—an experience which is somewhat startling to the tourist who ventures in one for the first time.

But the boat of greatest importance and interest is the kayak, which is certainly one of the most marvellous adaptations of natural forces to human use which has ever been made. This remarkable boat is a logical but most ingenious evolution from the birch-bark canoe of the northern Indian tribes of America. The various stages of development from the light and open boat of the northwestern tribes to the closed and water-tight shell of the kayak of Greenland can readily be traced around the shores of Alaska and British America.

The framework of the kayak consists of wood or bone, and is about twenty feet long, pointed at both ends, and about two and a half feet wide, and of the same depth in the middle. Over the frame there is tightly stretched and closely sewed a covering of tanned sealskin, impervious to water. The whole thing is so light that a boy of twelve can carry it with little effort. In the middle of the top there is a hole just large enough to permit the owner to insert his body, so that he can sit on the bottom and stretch his legs in front of him. Sealskin mittens and a sealskin coat, with a hood for the head and a rim for a close-fitting attachment to a corresponding rim around the aperture in the kayak, complete the protection from water. Thus fastened in, and sitting bolt upright, and provided with a paddle

Fig. 32.—Greenland kayaks.

flaring at both ends, the native can defy the winds and waves which would swamp an ordinary boat.

The first sight of a kayak in Greenland waters is exciting in the extreme. It is likely to occur, as with us, when lying off an unknown harbour waiting for a pilot. The whistle has been blown long and loud, the cannon has been fired and the rocket discharged, and still no response from the shore. At length, when patience is almost exhausted, there appear three or four black specks on the top of the distant swells of the broad bay, how

FIG. 33.—Kayakers coming out to meet us.

distant we are little prepared to estimate because of the excessive clearness of the atmosphere. As they get nearer we begin to see a curious motion somewhat resembling the arms of a windmill. These are the kayaks with their several occupants striving to outstrip each other in a race for the coveted job of conducting the ship into harbour. Already they are far ahead of the larger boat, which comes lagging along in the rear.

On reaching the ship the most fortunate kayaker loosens his coat from the rim of his boat, of which he seems to form a part, with much difficulty wriggles himself free from its entanglement, and is brought on board. Those who fell behind in the race soon arrive, and loiter around the ship, some of them resting quietly like ducks upon the water, only occasionally dipping down one end of their curious paddles to resist the force of some unusual wave, while others display both their own skill and the capacity of their kayaks by various manœuvres which never fail to astonish spectators. Now one will perform a somersault, or a series of somersaults, with his kayak, or will dart forward like lightning at right angles to another kayak and jump completely over the bow of it. To the natives the motions necessary to preserve equilibrium are almost a second nature, having become instinctive from childhood. But unfortunate indeed is the European who attempts any antics in, or even ventures into, one of these boats. The adult who has not already learned its management had better not attempt to learn.

At one time, while in camp at Ikamiut, when the wind was blowing a gale, shutting us up all day in our tent and tossing the waves of the fiord into such commotion that it would have been madness for any large boat to have ventured upon the water, we were thrilled by the cry that some kayakers were coming. They were three that belonged to the little settlement, and had come that day, as a matter of course, from Sukkertoppen, which was twenty miles distant. Upon reaching the shore and pulling themselves loose from their shells, the kayakers ran their hands into the apertures from which they had drawn their limbs, and brought out various objects of merchandise which they had pur-

11

chased at the store for their families. Then they severally took up their kayaks and carried them to a secure
place, and disappeared in the igloos, where their families
soon joined them to talk over the adventures of the
week. To us they seemed like inhabitants of the sea,
who were accustomed to shed their skins on coming out
of the water.

The kayak is equipped with various implements of
the chase. First, there is a bird spear, consisting of a
short handle of wood pointed with bone and surrounded
a little way back from the point by a circle of barbed
bone lance heads projecting forward, designed to give a
whirling motion to the missile, and, in case the object is
not squarely hit by the point, to make more sure of its
entanglement. Then there is the harpoon for the seal,
which is arranged with a loose joint, so that after the
spearhead has penetrated the animal it becomes detached from the shaft and remains connected with a
thong in the hands of the hunter. This thong is also
attached to a float, consisting of the inflated skin of
some large animal, sewed together so as to be air-tight.
This is thrown into the water and prevents the escape
or sinking of the wounded animal. Various kinds of
fishing tackle are also natural attachments of the kayak
when fully equipped. At Ikamiut, all these implements
were of native manufacture.

The houses of the Eskimo are as admirably adjusted
for their winter life as they are ill-adapted for summer
residence. In the far north they are built of snow, but
in southern Greenland of stone and turf, the walls rising
from three to five feet above the level of the ground,
while the roof is nearly flat or only slightly arched, and
covered with sod. One might easily pass by an Eskimo
settlement at a little distance and be unable to distin-

guish it from the clusters of natural mounds which occasionally occur.

FIG. 34.—Kayakers throwing bird spears in the harbour of Sukkertoppen.

The snow houses are, of course, always temporary structures, but they can be erected in a very short time; indeed, almost as quickly as a tent can be set up. Schwatka describes in a very interesting manner the course of the natives upon reaching a suitable camping place, when their first movement is to go around thrusting their spears into the snowbanks to find one sufficiently deep and dense for their purposes. On finding such a bank, they exclaim, " Plenty warm ! plenty warm ! " and forthwith proceed to carve out blocks from it, which are soon built into a dome-shaped hut fully justifying the exclamation, and which will defy the strongest and the coldest winds which even the arctic zones can furnish. These snow huts have to be abandoned for skin tents upon the approach of summer, which is duly announced to the inmates by the falling in of the roof.

The interior structure of the Eskimo house is extremely simple. The single room consists of an oblong quadrangle, upon one side of which is a shelf of boards about a foot from the floor, and wide enough for the adult members of the family to stretch themselves upon it at full length, with their feet to the wall. This is their sleeping and lounging place, occupied both day and night. A low partition in this shelf separates one family from another, while a narrow passageway extends along the side of the room through the entire length. At the farther end of this passageway, and at the end of each partition, are found the native lamps, around which the women habitually gather, and where they sit and jabber and sew and complete the process of tanning bird skins and seal skins by chewing them faithfully with their well-preserved teeth, adding variety to the scene by pausing now and then to trim their lamp—a process requiring considerable skill. The lamp consists of a shallow basin hollowed out of a piece of soapstone, in which there is a quantity of seal blubber. The wick is formed from a species of moss, which by frequent renewals is made to serve the purpose very well. To start

FIG. 35.—The better class of Eskimo houses at Sukkertoppen.

the lamp, some oil is expressed from a piece of blubber in the mouth and spurted into the hollow containing the general mass; but when once lighted, the heat of the flame melts the blubber and keeps up the supply of liquid fuel, very much as it does in an ordinary candle.

Light from the outside is admitted into the igloo through a single window facing the south. In primitive times, and still in the more remote settlements, the window is made of the translucent entrails of the seal, but now in the principal settlements glass is used. The sun is absent for so much of the year, however, that windows are of little account, while during the season of almost perpetual sunshine their winter habitations are deserted for tents. Of necessity, therefore, artificial light is their main dependence indoors, and their lamps are kept continually burning.

The entrance both to the ordinary igloos and to those made with snow, while designed to permit the passage of the inhabitants to and fro, is also planned to permit only so much circulation of air as shall provide ventilation and lower the temperature as little as possible. This passageway is narrow and crooked, so that, to enter, one has to get down upon his hands and knees, or be very deft in maintaining his balance in a stooping position. There are current numerous stories of fat men being wedged into the passageway to another man's igloo, so that they have had to be extricated by force, or relieved by tearing down the house. It is related of a very tall German missionary, that upon making his pastoral visits upon the native families, the only way of obtaining access to their igloos was by snugly incasing himself in a sleeping bag, and asking his parishioners to haul him in and out, as they would the carcass of a seal.

During the midnight storms of the long winter the enveloping snows afford additional protection to the inmates. Indeed, so well protected from the cold are these igloos in southern Greenland, that, without any fire except what is furnished by their lamps, the temperature is kept up to a high degree of warmth ; so that, for comfort, clothing is pretty much discarded by the natives when once the house is entered. This practice secures both present comfort and the thorough airing and drying of their clothes, which is an important sanitary result.

Much is said about the filthiness of the Eskimos, and, indeed, nothing can be more repulsive than the surroundings of an igloo in the early summer, after the sun has melted away the enveloping snow and laid bare the accumulated refuse of the winter. But it should be remembered that, so long as this remained frozen and covered with snow it was inoffensive and unobserved, and that ordinarily the Eskimo method of house-cleaning is very effective, which is to abandon the igloo upon the approach of summer, and suffer the elements to have free sway until the approach of winter, when the building is reconstructed after its long airing, and its inmates are securely buried up again.

So long as the natives were able to live in tents during the summer, it can not be said that their sanitary condition was unfavourable; but in these later days of contact with European civilization, when they have been tempted and enabled to kill the reindeer in excessive numbers for purposes of trade, they have so diminished the supply of skins that now they are unable to provide themselves with tents as formerly, and are often compelled to live in their igloos during the entire year, and so are prevented from moving from place to place as freely as in earlier times. Under such circumstances it must be confessed that the sanitary conditions are often deplorable, and this is showing its effect in the increasing prevalence of consumption and various other diseases.

The supposed bad effect on individual morality of such community of residence as is found among the Eskimos is by no means so great as one would at first imagine; for only the parents, the girls, and the smaller children sleep closely huddled together on the broad shelf, while the unmarried men lie by themselves on narrow shelves especially prepared for them, and the very crowding of the room secures for the girls the continual protection of a chaperon without her presence being obtrusive. Moreover, the difficulty, if not the impossibility, of a woman's supporting herself and a child without a husband and a kayak, renders illegitimacy a most serious calamity, and secures a pretty general observance of marriage rites and vows.

Such is the rigour of their conditions of life that the hardships of the unfortunate among the heathen Eskimos are to us almost inconceivable. According to Dr. Cook, all superfluous individuals in the far north are remorselessly permitted to perish. When a woman is about to give birth to a child, she is put in a house and given frozen meat sufficient to last for two weeks, and also some blubber and oil. If she survives the ordeal, and the baby is heard to cry, the others will come in and help her; but if the baby's cries are not heard, the house will not be entered again. Mothers nurse their children for four or five years; but if the mother dies before the child is three years old, the child is left to perish with her; while if the father dies before the child is three years old, the child must be killed, or the mother will

Fig. 36.—The Catechist's daughter in full dress.

lose all hope of obtaining another husband to support her. In times of famine the childless women and the old women are turned out to starve, and cannibalism is resorted to for self-preservation.

Almost all writers speak of the Eskimos as deteriorating in physical characteristics, and as rapidly approaching extinction. Doubtless there is much truth in this view, for there can be little question that contact with the Europeans has had an injurious effect upon the race in various ways. Indeed, trade with civilized nations has been almost necessarily detrimental to the Eskimos ; for they have little sense of the relative value of things, and are so inexperienced and improvident that in barter they are sure to get the worst of the bargain, and to exchange articles of prime necessity to

Fig. 37.—A typical Eskimo boy.

themselves for those which are of scarcely any real value to them.

There can be little doubt that if the Eskimos had been left to the tender mercy of traders they would

long ago have become extinct. Especially is the craving
of the people for stimulants such that they will part with
almost everything they have for coffee, tobacco, and alcohol. Through the wise precautions of the Danish Government, however, the sale of alcohol is absolutely pro-
hibited, so that natives are saved from its evil effects,
while the sale of tobacco and coffee is so limited by various regulations that the people are not allowed to im-
poverish themselves too much in obtaining them.

Even the use of firearms has been of doubtful advantage to the Eskimo, for, while it has enabled him to
kill more game of certain kinds, it has led to much
lamentable waste. Through the facilities which have
been furnished by firearms for capturing reindeer, there
has been imminent danger of the extinction of this important animal in Greenland, and the number has actually been very greatly reduced. From 1845 to 1849
twenty-five thousand reindeer were annually killed in
Danish Greenland, and sixteen thousand skins per annum were exported; but since that time the numbers
have been so diminished that from 1868 to 1872 not
more than one thousand per annum were killed, and
the export of skins had absolutely ceased. Consequently,
reindeer meat is no longer in daily use, and the reindeer
skins, so essential for the protection of the inhabitants
against the cold, are not sufficient to supply the home
demand.

In hunting the seal the use of firearms involves the
loss of many animals, which sink after the shot before
they can be captured; and the gun is so awkward a
weapon to carry in a kayak that in many situations the
Greenlander is hindered, rather than helped by its aid,
so that the chances of securing a seal by the use of his
harpoon and float are about as good as they would be

with the rifle. When in quest of birds, the noise of the gun frightens the whole flock and makes them wary; while with his bird spear the kayaker can approach the rookery or the flock as it floats upon the water and quietly capture as many as his small boat can carry, without alarming the great numbers which are unharmed. Guns are said to be of no avail in hunting the walrus, because his skull is so thick that a rifle ball will not penetrate it.

The tendency of civilization to diminish the size of the families occupying one house is also generally regarded as deleterious and productive of poverty. In 1870 there were only one hundred and five houses in Danish Greenland having more than sixteen inmates each. At that time the highest number in any house was thirty-six. Formerly the average was probably as high as forty.

Serious famines have occurred at various times during the present century, while epidemics have thinned the population to an alarming degree. Still the total number has not diminished during the last forty years, but, on the contrary, has slightly increased. While in 1855 there were but 9,644 natives on the west coast, in 1889 there were 10,177. Hans Egede, one hundred and fifty years ago, estimated the population at 30,000, but this estimate is probably a great exaggeration. At any rate it is not based upon anything like an accurate census, such as is now taken from year to year.

Two extracts from Graah's interesting narrative will give a better idea both of the original characteristics of the Eskimos and of their social customs than any generalizations can do. We may premise, however, that originally among them the winning of a wife proceeded upon the supposition that she was unwilling to enter into that relation, and so had actually to be forced into it. Many incidents are related of the persistence with which the woman has sought to evade being captured by her lover. Foreigners who have ventured to interfere for the rescue of the young woman have usually found, however, that the unwillingness on the woman's part was not invincible, for she is sure to prefer the rule of her lover to the protection of her impolitic defender. It is a common remark, also, that the Eskimos generally refrain from expressing in a public manner their interest in one another. Both their greetings and their good-bys are as undemonstrative as possible. Christianity has modified rather than eradicated these characteristics. In Danish Greenland the marriages are performed by the clergyman on his occasional rounds. But when the crisis approaches it is usually difficult to find the groom, who would be ashamed to have it appear that it was a matter in which he was at all interested. Lovers, while in each other's company, show no interest in each other.

The first of the incidents which we quote from Graah occurred after he had been icebound seventeen days upon the eastern coast, and his provisions had run short, putting his Greenlanders much out of spirit. "For the rest," he says, "nothing of any moment occurred during our detention at this place except the killing of a bear, to which one of our kayakers had nearly fallen a victim. The poor fellow had been sleeping on the ground in the open country, and was awakened by the hard breathing of the animal close by him. Springing up, he escaped in his boat, and felled him from thence with his arrows. The Greenlander who met with this adventure, and whose name was Ningeoak, was a lively, merry fellow of some twenty years of age, very much

addicted to antic tricks—in fact, the clown of the party.
Like his fellows, each of whom had, on our setting out,
selected a helpmate for himself from among the women
of the party, he had made court to several, in succession,
of his fair countrywomen, but had been refused by all
of them on the plea of his being a ‘Nellursok’—that is
to say, a heathen, literally an ignoramus. More than
once he had begged me to make intercession with them
in his favour, but all my efforts had proved of no
avail. The dead bear proved, however, a more efficient
advocate, his conquest of it (for a successful bear hunter
is held in Greenland in high repute) making so deep
an impression upon them that I verily believe he might,
if he had chosen it, have had them all. Ningeoak’s
pride had, however, been deeply wounded by their previous rejection of his suit, and, to revenge himself, he
chose for his helpmate a superannuated beldame, the
ugliest of the whole party.” *

On a later occasion, after describing the collection
of two hundred natives for a festival and tamboureen
dance (which continued all night), but which he was
unable to attend on account of illness, Graah writes : †

“On waking this morning, I heard the tamboureen
of my Greenland friends still going. I made haste,
therefore, to join them, and though when I reached the
spot they were on the point of breaking up, they continued their dance a little longer on my account. To
form an adequate conception of the dance I witnessed
on this occasion it is absolutely necessary to have seen

---

* Graah’s Narrative of an Expedition to the East Coast of
Greenland, p. 72.

† The circumstances concerning this gathering have already
been referred to on page 120.

it. To describe it is no easy matter. The tamboureen,
as I have termed it, employed by them by way of musical accompaniment, is a simple ring, or hoop, of wood,
with a piece of old boat skin, well saturated with oil,
stretched tightly over it, and furnished with a handle;
this one of the party holds in his left hand, and, taking
his station in the centre, while the rest form a ring
about him, and throwing off his jacket, strikes with a
small wooden stick, extemporizing, after a brief prelude,
a song, the subject of which is the chase of the seal or
some other, to them, important incident or event, the
whole assembly joining at the end of every strophe in
the chorus of 'Eia-eia-a! Eia-eia-a!' During this per-
formance he makes unceasingly a sort of courtesying
motion, and writhes and twists his head and eyes in the
most laughable style imaginable. Nothing, however,
can equal in absurdity the movements of his nether
man, with which he describes entire circles, nay, figures
of eight.

"This tamboureen dance is in high esteem among
the Greenlanders. When about to take part in it, they
put on their best holiday apparel, and the women take
as much pride in performing it with what they consider
grace as our young belles in dancing a quadrille or a
galop. It serves, however, not merely the purpose of
amusement, but constitutes at the same time a sort of
forum, before which all transgressors of their laws and
customs are, in a manner, cited, and receive their
merited reproof. When a Greenlander, to wit, thinks
he has sustained a wrong or injury at another's hand,
he composes a satirical song, which all his friends
straightway learn by heart, and then makes known
among the inhabitants of the place his intention of
bringing the matter to arbitration. On the day ap-

pointed, the parties, with their partisans, assemble and
form the ring, which done, the plaintiff, singing and
dancing as above described, states the case, taking occasion to retaliate on his adversary by as much ridicule
and sarcasm as he can devise, to which, when he is
finished, the other, singing and dancing in his turn,
replies; and thus the case is pleaded, till both have
nothing more to say, on which the spectators pronounce
sentence at once, without appeal, and the adversaries
part as good friends as if nothing had happened to dis-
turb the harmony of their friendship. In this way the
debtor is sometimes reminded of his debt and the evil-
doer receives a merited rebuke for his misconduct. In
fact, a better system of prevention and punishment of
offences—one, at least, better adapted to the disposition
of the people among whom it obtains—could scarcely be
devised, as there is nothing of which the Greenlander is
so much afraid as to be despised or laughed at by his
countrymen. This apprehension, there can be no
doubt, deters many among them from the commission
of offences ; and it is to be regretted that the missionaries, losing sight of this peculiarity of their temper,
have abolished this national dance on the west coast." *

A brief account of two Sabbath services which I
attended will furnish some important colours in the
picture of existing life in Greenland.

In the early part of August, 1894, as already described,
I set out with a party of eight from Sukkertoppen to
spend a week in camping close by the projection of the
inland ice, which there comes down to the head of the
lonely and picturesque fiord of Ikamiut. A hard pull

* Graah's Narrative of an Expedition to the East Coast of
Greenland, pp. 107, 108.

at the oars for twenty miles brought us, late on Friday
night, to our destination. On Saturday the weather
was unpropitious. The wind blew hard, and the air was
full of fog and drizzly rain. We were able to do little
but sit in our tent and cultivate the acquaintance of our
strange but kindly and well-disposed neighbours. They
were curious to see everything we had, and to know
both what it was and what it was for. It must be confessed also that we were equally curious to learn everything about them. But, in token of their good will the
women and children brought us an abundant supply of
moss and crowberry vines with which to carpet our tent
and to disguise the hardness of our rocky floor.

Sunday morning came, and it was still cold and
rainy. While we were eating our breakfast and shiver-
ing over our coal-oil stove in the tent, a man of mild
appearance and diminutive stature came to the door
with a hymn book and a Bible in his hands, and pointed
to them to indicate, as we surmised, that there was to
be religious service somewhere in the settlement. But
he did not linger long, and so silently disappeared that
we did not see where he went, and hence were at a loss
to know where the service was to be held, for the settlement was squalid in the extreme, and no one of the
three igloos seemed better than the others.

But on going down to our boats we heard singing in
one of the igloos. Stooping down before the low door
and pushing it open, on our hands and knees, we were
welcomed by motions into the most interesting church
service I ever expect to attend. To our eyes the room
in itself was dreary beyond description. The low walls
of stone and turf were reeking with moisture, while water distilled freely from the sod roof in various places,
and, as one walked along the passageways, spurted up

FIG. 38.—Our camp at Ikamiut.

from the crevices between the loose stones with which the floor was covered. The only dry place was the shelf, elevated about a foot, on the north side of the room, which for the regular inmates was their sleeping place by night and their lounging place by day. A cylindrical sheet-iron stove near the door was now cold and lifeless, because the creeping vines and peat were so wet that it was impossible to kindle a fire. A lamp of seal oil freshly distilled from the raw blubber was burning in the other end of the room, being the special property and care of the oldest woman of the household. In no place could one stand erect.

Yet here was gathered the whole community for Sabbath morning worship. Of course, I could not understand the words of their hymn, but the tune was a grand German choral, in which all united, maintaining perfectly the slow, dignified, and effective movement. Then followed a sermon from the little man, who proved to be the catechist living in the place. This was delivered in the Eskimo language, and with eloquence and effect, though from the lowness of the room the speaker was compelled to remain in a sitting posture. The only intelligible words to me in either the sermon or the prayers were the amens, in which all joined. Finally the service was closed with another hymn sung to an equally impressive German choral.

Not to be excelled in respect paid for the Sabbath, we arranged an English service in our tent after the midday meal, and circulated the notice among the Eskimos in the same manner as that employed by the catechist in the morning. We should have been glad to have asked them into the tent, but, as they had not yet learned that cleanliness is next to godliness, prudence suggested that they be excluded; so I stood in the door of the tent with our own company massed near the entrance, while the Eskimos, notwithstanding the inclement weather, gathered in front of the tabernacle. But they were all there, listening with the utmost devotion to the singing and the service, of which they understood not a word. Such was our first Sabbath.

The second Sunday service was at Sukkertoppen in a tastily built church seating about four hundred. The

FIG. 39.—Church at Sukkertoppen. Men carrying a umiak in the foreground.

room was crowded to its utmost capacity—the men sitting upon one side and the women and children upon the other, while a score or more of dusky but bright-eyed babies peered over the shoulders of their mothers or older sisters and added to the singularity of the service by their gentle but constant crooning. Numbers also of the late-comers were sitting upon the floor in the back part of the church.

The services were conducted by a native catechist,

and were essentially the same as those which we had attended the week before. But there was here a reed organ, upon which a native played, supplying the interludes and harmonies that regularly accompany a German choral. Some years before I had been impressed with this form of church music in the grand Cathedral of Cologne, but had supposed that its power was largely due to the exceptionally favourable conditions under which it was there rendered; the powerful organ, the immense congregation, the hearty enthusiasm born of centuries of practice, all seemed to be unique at Cologne. But here, amid Greenland's icy mountains, was a recently converted heathen people singing their praises to God in the same noble harmonies that satisfy the congregations of the most highly cultivated musical people in the world—and singing them, too, with equal perfection and enthusiasm. Probably each individual voice alone would have sounded execrably, but all together blended into a dignified volume of sound of the noblest and most satisfactory character. I had come to Greenland to learn effectually a most important lesson in church music. The reports of other parties respecting the Sabbath services at other places are to the same effect, and show that our experience was not exceptional.

A superficial glance at the condition of woman among the Eskimos leaves upon the ordinary European the impression that she is very much oppressed, and is little less than a slave to the men. More careful inquiry, however, will show that this is far from being the case; for, in fact, the women are exceedingly cheerful, and in the division of labour have reason to be, as they are, well satisfied with the part assigned to them.

When a seal is towed to land by a kayaker, it seems, according to the standards of American civilization, an

extremely selfish and lazy act for the kayaker to abandon it to the women and children, leaving them the
"drudgery" of all the further work, while he takes
his kayak under his arm and retires to the igloo, or
stretches himself in the sunshine upon some soft bed of
moss. But really this is the proper thing for him to
do, for the whole existence, not to say comfort, of the

FIG. 40.—A company of Eskimo women on the outlook.

family depends upon the man's success in capturing the
wild animals upon which they subsist; and this is by far
the most difficult and exhausting part of the work of
maintaining existence amid the conditions of life which
there prevail. For days together, perhaps, the hardy
kayaker has faced the fiercest elements without shelter
and with little food, and with the thermometer far below freezing. Everything depends upon his success in
the chase, and the heroism and patience displayed by
the hunters is almost incredible, while the physical

exhaustion often proceeds to the very limit of endurance, and the accidents to which they are exposed are
numerous—about ten per cent of the deaths being from
that cause. Of four thousand four hundred and seventy
deaths recorded in one portion of southern Greenland
from 1782 to 1853, four hundred and fifteen arose from
accidents to kayaks. In fact, the whole life of the successful Eskimo man is one of daring and heroism, such
as is well calculated to enlist the admiration of woman's
heart; so that the part of the work which the women
undertake to do is performed with great alacrity and
cheerfulness, and is such as in the main could not well
be done by the men without interfering with their effi-
ciency in the more imperative duties which devolve upon
them. The women, being always at home, naturally
have charge of the material which is brought in for the
support of all.

It is a busy scene which follows when a kayaker has
captured and successfully brought to land a seal of ordinary dimensions. The women immediately seize the animal, and, stretching it out upon the shore, strip it of its
skin, which in due time they are to manipulate for the innumerable uses to which it may be put. Chunks of blubber are distributed around to the eager crowd of children
and others, who devour them with all the avidity which an
American schoolgirl manifests for candy. What remains
of the blubber is carefully stowed away for use in the
ever-burning lamp. The meat also is stripped off from
the bones in preparation for a universal feast, and is in
general eaten without being cooked; while the undi-
gested food in the stomach of the animal is preserved
as a most precious delicacy, and the sinews and entrails
are laid aside to be manufactured into thread and thongs
of most enduring character.

The preparation of the skins for use varies according
to the objects in view; but one of the most essential
operations, which is always performed by the women,
consists in chewing the hide—a process which serves
the double purpose of removing from it the adhering
fat and oil and of rendering it pliable. Chewing is
done by doubling up the hide into a fold, so that a portion can readily be inserted in the mouth, where it is
munched until all the fat is extracted and the whole
made soft. This work renders it necessary for the
women to have good teeth, and it is noticeable that
Providence has in general supplied them with such;
whether as a direct result of their use from childhood,
or by the indirect process of natural selection, it is difficult to tell. But it is certain that women provided
with teeth well adapted for the preparation of hides
must have a great advantage over those less generously
provided, and must be sought for more frequently as
suitable companions for the most heroic and successful
hunters.

The skins of the eider duck, the kittiwake, ptarmigan, and other birds are prepared by the women in a
similar manner, this serving as a sort of knitting-work
in the intervals between the severer tasks of rendering
pliable the skins of the larger animals. So severe is the
work of munching the seal skins that a day of labour
upon them is followed by a day of rest. The women
are also relied upon to mend the seal-skin boots, which
are so essential in all Greenland expeditions. These,
too, have to be rendered pliable by chewing before the
rips are closed up or the patches sewed on. It is evident
that the Greenland women get along very comfortably
without the use of gum.

The women show much taste and skill in the manu-

facture and ornamentation of their clothing. The leath-
er for their boots is usually coloured a beautiful blue,
and the boot-tops are ornamented with stripes pleasing
both in colour and in form. The short trousers are
made of undressed reindeer skin, variegated with stripes

Fig. 41.—Eskimo household servants. Married and unmarried.

of fur from other animals of a different colour; and in
the summer time the blouse, which comes down to the
hips, is covered upon the outside with cloth of appropriate figure, and ornamented with a band of a different
colour around the bottom; while a necklace of coloured
beads covers the shoulders and breast when in full dress,

and the hair is pulled up from every side to the top of the head, and, after being made into a compact roll, is tightly bound with a ribbon, whose colour indicates the condition of the wearer. The unmarried women bind their hair in red, the married in blue, and the widows in black, mingling blue with it, however, and surrounding the forehead by a narrow band of white when they have mourned a sufficient length of time and are ready to marry again.

The great abundance of birds supply both men and women with ample material with which to provide their blouses with a warm lining. The skins are stripped bodily from the breasts of the birds, and, after being carefully chewed and tanned, are sewed together with much skill to fit the body, with the feathers inward. Beautiful quilts are also made from the skins, presenting upon one side the soft down of the eider duck, and upon the other the thicker but coarser feathers of the kittiwake, the auk, or the ptarmigan.

The women, like the men in Danish Greenland, can all read and write, and are extremely fond of music and dancing. In the long days of summer they never seem too tired to gather on some level place in front of their houses and dance their beautiful quadrilles to music of their own singing. In one favourite dance the partners, at certain intervals, draw off from each other in a very graceful manner, and shake the forefinger of each hand in succession three times in the face of the other, and clap their hands together three times in the rhythm of the music, when they join hands again to move on as before. Indeed, the heart seems as merry in the midst of Greenland's icy mountains as it is in the sunny islands of the south Pacific, proving how easy it is everywhere for man to rise above his environment and make his life worth living.

Doubtless Christianity has done much to ameliorate the condition of women in Danish Greenland, but even among the heathen Eskimos on the east coast and north of Melville Bay the apparent hardships to which widows and orphans are subjected have some foundation of excuse in the severity of the conditions of life to which they are compelled to adjust themselves; for so near are they living all the while upon the verge of starvation that the community can not take upon itself the responsibility of providing for an indefinite number of non-producers. Each man has all he can do to provide for his own family; so that when a widow is left with infant children there is an appearance of hard necessity driving them to leave these helpless ones to their fate. It is often a question between leaving one family to starve by itself or all starving together. The problem and the extremity of the resort are not so different as they seem from those which sometimes confront us in civilized lands, where the population in any district has increased faster than the means of subsistence have done.

CHAPTER VII.

THE discovery of Greenland by Europeans took place in the latter part of the tenth century, when colonists from Iceland and Scandinavia were attracted thither in considerable numbers and established a civilization which maintained itself for four centuries. The discoverer was an Icelander named Eric the Red, a freebooter, who was compelled to leave his country for his country's good. A few years before he set out upon his voyage of discovery another Icelandic adventurer, named Gunnibjorn, had been driven by fierce storms westward into unknown waters, where he came in sight of islands which probably belonged to the eastern coast of Greenland, or perhaps of the coast itself, but he made no attempts to land. His reports, however, stimulated the hope of Eric, and guided him in his efforts to discover a new country where he should be undisturbed by enemies. He was successful in finding harbours in the southern part of Greenland, where he spent two winters. This was about the year 982, and the portion of the coast explored by him is that between Cape Farewell and Julianshaab, about latitude 60°, the region known as the Eastern Bygd, or Colony.

When the term of Eric's banishment had expired he went back to Iceland, and, partly by giving an attractive name to the region, induced twenty-five ships to return

with him ; but, of these, eleven were driven back or
wrecked. Rink expresses the feeling of all who read
the history, when he says that it is amazing " that Eric
and his followers in the course of three or four years

Fig. 42.—Arrival of the Rigel for relief of the Miranda.

should be able to find out what in modern times only
repeated exploring expeditions extending over more than
one hundred years at length succeeded in rediscovering,
notwithstanding the enormous advantages furnished by
science and new inventions and the assistance rendered
by government to later explorers.   The first obstacle encountered by the voyagers to Greenland is the border of
drift ice that commonly guards the coast, which is generally so narrow that the vessels outside can sight the
land across it, and yet so thick and dense that it becomes more or less impossible to force a way through
it.   Second, if we succeed in reaching the shore we
meet with labyrinths of rocks, islands, and bluff head-

lands, all of them barren and desolate, offering at first sight nothing that would seem to secure the necessary means of subsistence for human inhabitants. Numerous inlets are found between these islands and peninsulas leading farther into the inner country, and here, no less than thirty to forty miles' distance or even farther away from the outer islands, small tracts of flat lands will be found which might well have appeared inhabitable to the ancient discoverers. How they ever found out these spots, far apart from one another and totally hidden behind the craggy and ice-covered highlands, is a matter of surprise. When Eric had first sighted the land opposite to the coast of Iceland he found it encumbered with the same barrier of drift ice which in modern times has frustrated so many hazardous attempts to reach the coast. We may also conclude that he attempted to force his way through it in many places before he sailed four hundred miles to the southwest and rounded Cape Farewell, and, having finally succeeded in landing, he must afterward be supposed to have tried hundreds of sounds, deceptive inlets and creeks, or *culs-de-sac*, on the coast bordering Davis Strait at a stretch of four hundred miles.

"With all the detailed knowledge which has now been acquired regarding the same course, the Danes having had establishments there for more than a century, no better localities for settlers have yet been found than those which Eric of old pointed out to his followers in the fourteen ships which succeeded in reaching the country." *

Three years later, in the year 999, Leif, a son of Eric the Red, visited Norway, where he met Olaf Tryggvas-

---

* Danish Greenland, pp. 4, 5.

son, the king, who was just then exhibiting the zeal of
a new convert in the spread of Christianity. By the
arguments of the king the adventurer was converted,
and in the following summer returned to Greenland,
taking with him a priest, whose persuasive arguments,
joined to those of Leif, were successful in the conversion of Eric, who, with all his followers, was soon
after baptized.

In due time the colonies spread as far north as Godthaab, in latitude 64°. But the only means of communication between the two settlements was by boat, and
six days were required for the passage. The east settlement eventually grew to contain twelve parishes and one
hundred and ninety farms, and supported two monas-
teries and a cathedral. The west district had four parishes and one hundred farms. Numerous ruins still
exist marking the site of these early settlements, one of
which is believed by Rink to be the very house in which
Eric took up his first abode. This, as indicated in an
earlier chapter, is situated near Igaliko, in latitude 61°
50', at the end of a fiord extending forty or fifty miles
toward the inland ice from Julianshaab. The steep face
of a precipice formed one side of the house, while the
walls on the other side and at the ends consist of red
sandstone blocks more than four feet thick. Until recently the entrance to this was still preserved, "meas-
uring about six feet in height and four feet in breadth."
Some of the stones in the wall measure from four to six
feet in cubical dimensions. The house was about forty
feet in length by twenty feet in breadth. The ruins of a
church are found near by, with fragments of sculptured
stone, indicating Christian burial places.

At this period Greenland was not wholly devoid of
literature. In connection with the establishment of

Christianity poets and annalists arose who have transmitted to us thrilling stories concerning adventurers, who, not satisfied with the laurels already won, pushed farther out to the west and south in search of new lands. According to these sagas, Leif succeeded in crossing Davis Strait and reaching the coast of Labrador or Newfoundland, and even extended his voyage as far south as Massachusetts, where wild grapes were discovered. Bringing back a cargo of timber and grapes to Greenland, great excitement was produced, so that in the coming year Leif's brother Thorvald, with thirty men, visited Vinland, where he spent two winters, but was eventually killed by the natives, who, from the description, are thought by many to have been Eskimos.

Two years later, after the return of Thorvald's expedition, Torfin Karlsefne came over from Norway and spent a winter in Greenland, and married a relation of Eric's. The following summer he sailed with one hundred and sixty men and reached Vinland, where he remained five or six years, having more or less intercourse with the natives. There seems little doubt that Newfoundland, Nova Scotia, and Massachusetts were visited by these adventurers, and that some of the people they encountered were also Eskimos. But no permanent settlement was made in the newly discovered land, and visits to North America became more and more infrequent, until they were finally discontinued entirely.

Meanwhile the coast of Greenland, both to the east and to the north, were explored, and settlements were made upon the west coast as far up as the sixty-fifth degree of latitude. A runic stone, indicating at least a temporary visit of these early colonists of Greenland, was found as far north as 73°. The date upon it was 1235, while there is some evidence that in 1266 an expedition visited Lancaster Sound and reached a latitude of 75° 46′.

From the number of Norse ruins existing in Greenland it is thought by some that the population may have reached several thousand, and, according to the records, while they had to import all timber and iron, their exports show that they raised sheep and oxen to a limited extent, and were successful in killing seal and walrus. Bread, however, was said to have been, then as now, utterly unknown, while reindeer, whales, and bears contributed to their necessities and comfort. Evidently the conditions of the country have changed but little since. It is said that as late as 1484 there were "forty men in Bergen, Norway, who were acquainted with Greenland navigation, and used annually to bring home precious cargoes from that country." After that date all communication with Greenland ceased until it was rediscovered by Davis in 1585.

Upon the rediscovery of Greenland vigorous efforts were put forth to find the descendants of the colonists who had been so long neglected ; but none have ever been found, and there has been great disappointment as to the extent of the ruins of the former settlements and of the cultivable land surrounding them. The fact seems to be that there never was much cultivation of the soil in Greenland.

The failure to find these colonies on the accessible shore of western Greenland has led to many efforts to find them on the east coast; but these have only resulted in proving that the east coast was never inhabited by Europeans. Even as late as 1883, Nordenskjöld cherished the hope that a habitable region existed in the interior of Greenland which he supposed was not reached by the heavy snowstorms that characterized the coasts. But the exploration of the interior has demonstrated that it is entirely enveloped with ice, and fur-

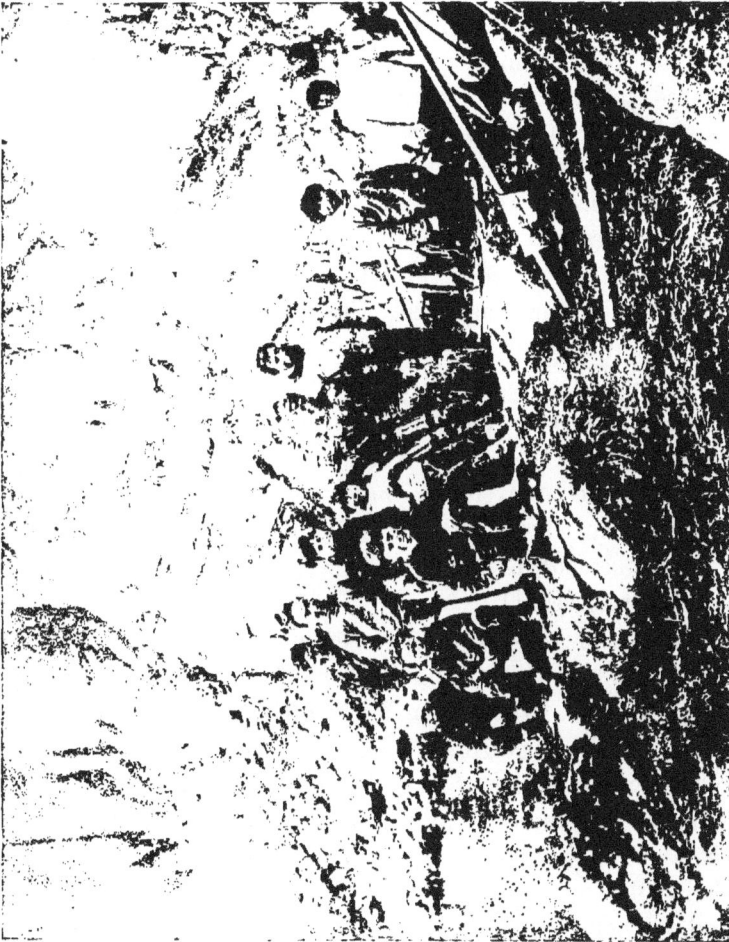

Fig. 43.— Eskimo family, showing Danish blood.

nishes no conditions for supporting the life of either man or beast.

A popular impression has been that the early colonies were destroyed by the Eskimos near the middle of the

13

fourteenth century, about which time it is said they first
came in contact with them in the neighbourhood of
Disco Island. But it is doubtful if there is any truth
in this legend. It is more likely that the decay of the
colonies was hastened by the " black death," which dev-
astated so much of Europe during the middle of the
fourteenth century, and which was extremely fatal in
Trondhjem, the Norwegian port which had most trade
with Greenland. Rink is of the opinion that the European settlers became intermixed with the Eskimos, who
advanced upon them in the latter part of the fourteenth
century, and that very likely their descendants are still
to be found among the present natives of Greenland,
many of whom even on the east coast show distinct
traces of European descent. As contributory causes of
the decline of the early Norse colonies, we must consider also the many feuds that were continually arising
among themselves, and the difficulty of maintaining a
comfortable subsistence when intercourse with Iceland
and Norway ceased.

After the rediscovery of Greenland by Davis, in 1585,
more than a century elapsed before effectual colonization
began again. In 1605 the King of Denmark sent out
an exploring expedition, which went as far north as the
sixty-seventh degree of latitude, and returned with many
articles of commerce and with two or three of the natives whom they had captured and brought away against
their will; but subsequent efforts were not so successful,
and the country was only infrequently visited during the
remainder of the century.

The present development of Danish occupation and
control dates from July 3, 1721, when Hans Egede with
his wife and children landed at the place now called
Godthaab, and established a Christian mission for the

conversion of the natives. It was hoped that the trade of the colony might be sufficient to make it self-supporting, but this proved delusive. Among the most interesting but unprofitable efforts of the Danish Government to infuse life into the colony was that of 1728, when a military expedition, consisting of twenty-five soldiers with their proper officers and eleven horses, was sent from Denmark to explore the interior. But the horses were of course useless and soon died, while the governor found that the interior was but a barren waste of ice. During the first winter the soldiers and other colonists who had been sent with them endured unspeakable hardships, and no less than forty died of diseases.

The difficulties encountered by Egede were such as would have discouraged any less devoted man. The language was difficult to master, and for years he had to depend on pictures to convey to the Eskimo the new ideas of the Christian religion. At first, also, they took him to be an *angekok*, or medicine man, whose authority could be sustained only by a miraculous healing of the sick. Two orphan boys to whom he had given special instruction left him at length, declaring that he was of no account, and could do nothing " but look in a book and scrawl with a feather," whereas their own countrymen " could hunt seals and shoot birds."

In 1731 the death of King Frederick IV, who had been the chief patron of the mission, brought things to a crisis, and orders were given for the Europeans all to come home. But Egede, who had just mastered the language and won his first converts, persuaded a few men to stay with him, and wrote a letter begging the new king not to abandon the work which had been begun with so much labour. This letter was effective, and the mission was continued.

Soon after this, in 1733, the Moravians established a mission at New Herrnhut, in close proximity to Egede. Since that time the Lutherans and Moravians have continued their work side by side. The Moravians now have six stations, but the main work has been done by the Lutheran missionaries on the foundations laid by Egede.

Egede remained in Greenland until 1736, when his wife died and he himself was taken ill, which compelled him to return to Denmark, where he spent the rest of his life in broken health. His son, however, remained behind, and his father's work was carried on with renewed zeal and increasing success until, at the present time, the Eskimos in Danish Greenland are all nominally Christians, about two thousand of them being connected with the Moravian mission and about eight thousand with the Lutheran Church. Eight Lutheran missionaries oversee the work of the numerous native catechists and schoolmasters who are employed to impart instruction directly to the people. These Danish missionaries usually remain in Greenland only about ten years each.

Notwithstanding the great difficulty of furnishing instruction to a people so widely scattered as the Greenlanders are, the rudiments of education have been so effectively imparted to them that nearly all can now both read and write. A printing office is maintained at Godthaab, where about fifty volumes have been published. These comprise the Bible, about twenty religious books, and sixteen entertaining story books, besides numerous schoolbooks. According to Rink, whose long residence in Greenland makes his testimony unimpeachable, " the greater part of the inhabitants are able to read tolerably well out of every book in their own lan-

guage. . . . The art of reading is not only familiar in
every house, but reading also forms a favourite occupation. . . . Carrying on correspondence by letters has become pretty frequent between the natives of different
stations. . . . Moreover, the natives seem to be peculiarly talented as to acquiring a good hand in writing.
. . . Scarcely any country exists where children are so
ready to receive school instruction as Greenland ; it is
almost considered more a diversion than a duty." *

Everything in Greenland is under the control of the
Danish Government, the management being committed
to a board (the Kongelige Grönlandske Handel) residing
in Copenhagen. Danish Greenland itself is divided into
two inspectorates, the northern and the southern, of
which the respective capitals are Godhavn and Godthaab.
The two inspectors are commissioned with the responsibility of carrying out the regulations of the board, and
within the limits of those regulations have almost absolute power. A foreigner is not permitted to remain any
length of time in Greenland without the permission of
the inspector.

In each district (of which we have given detailed
account in a preceding chapter) there is a subordinate
governor responsible to the inspector, and under him
various post traders, though the trade is all made to
centre in the capitals of these districts, which are visited
annually by one or two Government vessels. Ordinarily,
as has appeared in the detailed account, the capital of
the district contains only three or four Danish dwell-
ings, a storehouse, and a blubber-boiling house, around
which are gathered the casks to hold the oil and the
barrels to contain the fish which are to be exported. In

* Danish Greenland, pp. 214–217.

some of the districts there is a clergyman, a teacher, a physician, and a Lutheran church, though in most of them the religious services are maintained by a native catechist, while the clergyman makes only occasional visits.

At Sukkertoppen the Danish colony consisted, in 1894, of the superintendent, Mr. Bistrup, and his assist-

FIG. 44.—The Danish ladies at Sukkertoppen.

ant, Mr. Baumann, and their wives and three children, and Miss Fausböll, a niece of the superintendent and daughter of the Professor of Sanscrit in Copenhagen University. Altogether, this little company of Danes lead a life which is by no means devoid of attraction. They are all highly educated, and speak and read English readily. Their house is commodious, and well protected against the weather. A two years' supply of provisions is always kept on hand, to guard against any failure in communication with the outside world. An ample library, both of Danish and English books, furnishes them reading material during the long winter night, while the piano and their own cultivated voices provide them with music according to their various moods of feeling. Mr. Bistrup has been seventeen years in Greenland service, and, according to their regulations, will, after eight years more, be permitted to retire and return to Denmark upon a pension. Miss Fausböll is simply spending two or three years for an outing with her uncle, having charge meanwhile of the instruction of the children of the two families. They all speak with enthusiasm of the life, though longings for the home land could not altogether be suppressed.

The salaries of the superintendents are fixed at £250 a year, besides residence, fuel, and attendants. The salary of the inspectors is £328, while that of the clerks and assistants is £106. Besides these officials there are nearly 200 Danish outpost traders, seamen, and mechanics employed in Greenland, distributed pretty equally among the colonies and receiving an annual salary of £25 each, and, in the case of the traders, a percentage on the trade.

The discomforts and inconveniences of official life in Greenland are manifest, but are endured with little complaint, all seeming to make the best of their circumstances and to minimize their trials. In the posts where there are no physicians (which constitute a majority) it is necessary to be sick when the physician makes his semiannual round, in order to reap the benefit of his services. One of the ladies of Sukkertoppen was planning, when we were there, to return to Copenhagen the next season to have a much-needed operation in dentistry performed. Happily, we had an accomplished dentist from New York with us, who was able to do the work and relieve her of the burden of the long journey and separation from her family. It is related of the superintendent at Upernivik that he had brought to him annually from Denmark a file of daily papers, which were turned over to his wife with orders that a copy should be brought him each morning at breakfast a year after date.

Though the Government is entirely in the hands of the Danish officials in Greenland, through the exertions of Dr. Rink, who was inspector of southern Greenland for twenty-five years or more, local councils were organized from the natives in every district. These consisted of delegates from every station " elected at the rate of about one representative to one hundred and twenty voters." This native council is invited to aid the regular officials, especially in distributing the surplus profits of trade and in general advice with reference to the welfare of the district. " It holds two sessions every year, and the discussions are entirely in the Eskimo language." They are the guardians of the poor, and both investigate and punish crimes and misdemeanours and divide inheritances. In cases of high misdemeanour they are permitted to inflict corporal punishment. But the character of the Eskimo is so peaceable that there is little work for this council to do aside from the distribution of relief to the needy. During the first ten or twelve years after their establishment in 1857, the following were the principal causes submitted to trial: "One single case of having in passion occasioned the death of a person, and another of openly threatening; five or six instances of grosser theft or cheating, and as many of concealment of birth and crimes relating to matrimony; every year a few petty

thefts, and instances of making use of the tools of others
without permission, or of like disorders; and several
trifling litigations." *

FIG. 45.—The governor's house at Sukkertoppen.

With the great preponderance of women among the
Eskimos consequent upon the hazardous nature of the
occupations of the men, it would seem as if the servant-girl
question might be easily solved in Greenland; but such,
we are informed, is not the case. The native women
are restive under the restraints and duties of civilized

* Robert Brown, in Encyclopædia Britannica.

housekeeping. So long as they can be persuaded to remain in service, their pleasant faces, mild manners, and generally docile spirit make them unusually acceptable servants. Dressed in their attractive native costume, they move quietly about the house, attending to their duties, unencumbered by the long skirts of female attire in civilized countries, and, to the stranger, add much picturesqueness to the dinner parties so lavishly given by the Danish families. But their hearts are not in this service. They still prefer to mingle with those of their own kind, and wander away during the summer months on camping excursions or join in the long winter evenings in the more agreeable tasks of chewing and tanning and working up the seal skins and bird skins brought into the igloos by the adventurous hunters of the other sex.

In the year 1870 there were "in the service of the Danish mission 53 appointed teachers, besides several other teachers classed as seal hunters or fishers. In the service of the royal trade were 12 outpost traders, 15 head men and boatswains, 14 carpenters and smiths, 19 coopers, 15 cooks, 54 sailors and labourers, besides 10 pensioners and 33 midwives; five officers were enumerated as natives, but three of them are more properly Europeans. In the same year the Europeans numbered 237, of whom 95 were engaged in the trade, 8 were Danish and 11 Moravian missionaries, and 38 lived at the cryolite mines; the rest were women and children."

As already said, the trade of Greenland is a Government monopoly, having been taken up in the year 1774, when it had ceased to be profitable as a private monopoly. Rink estimated that the earnings of the Eskimo families average £8 each "from the produce of the hunt sold to royal officials." According to Robert

Brown, during the twenty years from 1853 to 1872 the average annual exports of the material brought in by the natives "consisted of 1,185 tuns of oil, 35,439 seal skins, 1,436 fox skins, 41 bear skins, 881 waterproof jackets, 1,003 waterproof trousers, 3,533 pounds raw eider down, 6,900 pounds feathers, 2,300 pounds whalebone, 550 narwhal ivory, 87 walrus ivory, and 1,817 reindeer hides." During the period of 1870 to 1874 "the mean annual value of the products received from Greenland (exclusive of cryolite) was £45,600; that of the cargoes sent thither, £23,844; and the mean expenditure on the ships and crews, £8,897. . . . The average profit of the Greenland trade was, for the twenty-one years between 1853 and 1874, about £6,600 yearly. The capital sunk in the 'royal trade' is calculated at £64,426; and, taking the whole amount of net revenues from the present trade during the period from 1790 to 1875, the interest on the capital being subtracted, the director considers that £160,000 has been earned." According to the reports of 1892, the exports for that year were valued at £32,000, which shows a considerable falling off from the earlier average. The product of the cryolite mine at Ivigtut, which was first opened in 1857, amounted to 14,000 tons between that time and 1864. In that year the exports rose to a much higher figure, averaging a little over 10,000 tons annually for the next nine years. Since 1873, however, there has been a slight falling off, the yearly average being only 8,000 tons. The total value of the cryolite exported since 1871 has been about £450,000. From this the Danish Government has received a royalty of a little more than £100,000. Though the production is nearly as great now as ever, the substitution of bauxite for cryolite in the manufacture of aluminium seems to render the mines rela-

tively less important than they were formerly supposed
to be.

It thus appears that the present relation of Denmark to Greenland is supported more by motives of
humanity than by hope of profit, and the regulations of
the Government have chiefly in view the protection of
the native population, and all certainly bear marks of
this humanitarian air. The prices to be paid for European articles are fixed for every year, those current
being printed in Danish and Eskimo and distributed by
the Government. The traders sell European articles of
necessity to the natives at cost, and bread at somewhat
less than cost, while twenty per cent profit over the cost
price in Denmark is all that is allowed to be charged
on general merchandise. The price paid for native
products is about twenty-two per cent of the value in
European markets. One sixth of the price paid for the
native products is "devoted to the Greenlanders' public
fund, spent in public works, in charity, and on other
unforeseen contingencies," the expenditure of which is
under the control of the council above described.

The Danish Government has also done an important
work in surveying and mapping the west coast, thereby
adding largely to our knowledge of the geography of
the country. The results of this work and all that is
incidental to it, such as the collection of facts concerning the language, character, and history of the Eskimos,
the early European settlements in the country, and the
meteorology of the region, have been published in an
elegant series of annual monographs, partly in Danish
and partly in English, entitled Meddeleser om Grönland.

The prospect of any marked commercial or social
improvement, however, is not encouraging. The cap-

ture of the whale has almost wholly ceased, while the reindeer have so diminished that there are scarcely enough to supply the wants of the natives, and some of the favourite haunts of the eider duck have become almost depleted of their occupants by thoughtless efforts of the natives to increase unduly the annual returns. Happily, however, most of the haunts, both of birds and of fish, as well as of the seal, are so guarded by Nature that they can not well be seriously interfered with by any agencies that are likely to be employed. It is greatly to be hoped that profit enough may continue to attend Danish control in Greenland to make it and its protection to the natives as continuous as it is beneficent.

CHAPTER VIII.

THE PLANTS OF GREENLAND.

THE verdure which the old Norse discoverer commemorated in the name Greenland is scanty in comparison with that of all other inhabited countries. Mountains whose surface is mostly bare rock, usually of sombre gray or dark colours, form the Greenland coast throughout its extent of four thousand miles or more. But even the rocks, except in the most precipitous and bleak places, are overgrown with lichens, which likewise are usually of dark and dull-gray colours, strongly in contrast with the snow that is drifted deeply into the ravines and there endures nearly to the end of the short summer. On the higher slopes of many of the mountains the snow is never wholly melted away, and so forms small local icefields, from which glaciers descend in the valleys. In the general view from a ship passing along the coast, the dark gray of the rugged landscape and the white of the *névé* fields and local glaciers, and of the larger glaciers flowing down from the inland ice sheet, suggest that this is the last part of the world, excepting only the more completely ice-enveloped antarctic continent, to deserve its attractive name.

Nearer inspection of the land, however, gained by entering any of its long western fiords, reveals many beautiful bits of scenery, where, between the frowning mountain sides and the deep, riverlike, crooking fiord,

188

green patches of sedges and grasses are bedecked with hardy northern and arctic flowers. Increasing variety of the vegetation, and many shrubs or even dwarf trees, often forming considerable thickets along the ravines, are seen in proceeding toward the head of the fiord; the moist and cold sea winds on the outer coast giving a mean summer temperature several degrees lower than that of the more sheltered, less stormy, and more sunny inner fiord valleys.

During the two or three months favourable for the growth of vegetation the headlands and islands of the coast, wherever the rocks are covered by the thin glacial drift or decomposing rock in sufficient amount to provide a soil, bear a considerable variety of mosses, sedges, and very low, matted shrubs. The most conspicuous of the shrubby plants there and nearly everywhere throughout the land tracts of Greenland is the crowberry or curlew berry (*Empetrum nigrum*), whose juicy black fruit, matured by the early autumn frosts, is much enjoyed by the Eskimos, and which Nansen and his companions found in luxuriant abundance, and ate of very heartily on their reaching the land at the western base of the ice sheet late in September. Dr. Rink tells us that it is the only berry or fruit used by the natives, and that, except in marshy places, one can hardly cut out a sod anywhere without including roots and branches of this heathlike procumbent evergreen. He further writes :

In warm summers berries begin to turn black in the middle of July, but they are not edible before August. At the end of this month the night frosts prevent their decay, and berries may be found throughout the winter by scraping away the snow. They are perfectly fresh when they come to light by the melting of the snow in May and June. The abundance of these fruits in favourable years is really surprising. Even in 69° north latitude the

FIG. 46.—The north side of Sermersut, showing proximity of icefields and vegetation.

creeping branches may be seen so laden with fruit that they resemble bunches of grapes, and almost blacken the ground.*

About the mouths of the fiords, having a somewhat
more sheltered position than the outer islands and the
capes of the mainland, two species of willow (*Salix
arctica* and *S. glauca*) and the dwarf birch (*Betula
nana*) thrive, fixing their roots in the clefts of the rocks
and creeping along the ground, with stems sometimes
six to eight feet long and one to three inches in diameter, much knotted and twisted, and rising to maximum
heights of two or three feet. Among the flowers noted
by Rink as found there in bloom during July and August are the Lapland rhododendron, several others of
the heath family, numerous species of saxifrage, willow-
herb (*Epilobium*), lousewort (*Pedicularis*), and bell-
flower (*Campanula*), the narrow-leaved arnica, species
of cinquefoil (*Potentilla*), crowfoot (*Ranunculus*), and
others, rose-coloured, blue, purple, yellow, and white,
which beautifully diversify the prevalent brownish
green of the spots occupied with vegetation.

Near the heads of the fiords, from ten or twenty
miles to fifty miles or more from the outer coast, southern Greenland has thickets or copses of alder (*Alnus
repens*) extending northward some fifty miles beyond
Godthaab, and of a white birch (*Betula alpestris*), which
is only found south of Frederikshaab. The dwarf juniper (*Juniperus nana*), which is the only coniferous
species occurring in Greenland, has a wider range but
attains its maximum size in these sheltered valleys, its
prostrate stems having sometimes a thickness of five to
six inches, though commonly no more than two or three
inches.

* Danish Greenland, p. 88.

14

Besides the crowberry, the only other edible berries are the bog whortleberry or bilberry (*Vaccinium uliginosum*) and the cowberry (*Vaccinium Vitis-Idæa*), both of which grow also in northern Europe and extend southward in North America to the White Mountains and the north coast of Lake Superior. The cloudberry (*Rubus Chamæmorus*) is common near Godthaab, where, according to Rink, it may be seen flowering prettily even on the outer shores, but its fruit very rarely ripens. This species, which also is European, extends south to the White Mountains, and occurs locally on the coast of Maine.

There are several plants of which the flowers, buds, leaves, or roots are eaten by the Eskimos, generally in a raw state. The most prized is the *Archangelica officinalis*, somewhat resembling celery. It grows to a height of six feet in favourable spots of the fiord valleys, and extends northward to Disco Island. The young stalks are brittle and sweet, with a flavour like that of carrots. The nutritious lichen called Iceland moss (*Cetraria islandica*) is common, and has a wide range in both latitude and altitude, but attains its greatest abundance on the outlying southern islands. The chief resources of vegetable food, however, are marine, as noted by Dr. Rink :

Seaweeds may perhaps be considered the most important vegetable diet of the Eskimo, because they have in many cases saved people from death by starvation. The species most commonly eaten is *Alaria Pylaii*, closely allied to the edible "hen ware" or "bladderlocks" of Scotland, the *Surdluitsok* (i. e., without hollowness) of the Greenlander, which has a soft stalk as thick as that of an asparagus, and headed by a broad leaf. . . . Another kind, *Chorda filum*, the "sea laces" of the English fishermen, the *Augpilagtok* (i. e., red) of the Eskimo, is considered more delicate, but is less abundant. Both these kinds are also eaten when there is no lack of food, but there is a third sort, smaller in size and far more common, which is only resorted to in time of need.*

The chief fuel used by the Danish people in Greenland is the peaty turf, six to eight inches thick, formed commonly on sloping hillsides by the matted growth of plants. It is cut in proper size for use and thoroughly dried in the summer—a necessary exercise of foresight which mainly prevents its use by the Eskimos. Bog peat similar to that of temperate countries, but attaining a thickness of only one or two feet, also occurs occasionally in the glens or ravines tributary to the southern fiords as far northward as Godthaab (latitude 64°).

In addition to the species of phænogamous plants already mentioned, the following, as enumerated by Dr. Rink, are the most abundant on the turfy slopes: four grasses, *Alopecurus alpinus, Hierochloa alpina* (also occurring on the mountain tops of New England and New York), *Poa pratensis* (common southward to the central United States, where it is well known as the Kentucky blue grass), and *Festuca brevifolia*; eight sedges, *Carex Wormskjöldii, C. capitata, C. nardina, C. glareosa, C. Vahlii, C. hyperborea, C. rariflora,* and *C. supina*; two species of cotton grass, *Eriophorum Scheuchzeri* and *E. angustifolium*; three species of wood rush, *Luzula spicata, L. multiflora,* and *L. parviflora,* of which the first and third occur on the summits of the White Mountains in New Hampshire; *Juncus biglumis; Salix herbacea,* a very small, herblike willow, rising only a few inches above the ground, found also on the White Mountains; *Thymus serpyllum*; *Diapensia Lapponica,* found on the tops of the White and Adirondack Mountains; four

---

* *Op. cit.,* p. 90.

of the heath family, namely, *Azalea* (*Loiseleuria*) *procumbens*, which again is a species of the White Mountains, *Ledum Grœnlandicum*, *Phyllodoce cœrulea*, and *Andromeda tetragona*; six species of saxifrage, namely, *Saxifraga stellaris*, *S. nivalis*, *S. rivularis*, *S. cœspitosa*, *S. tricuspidata*, and *S. oppositifolia*; and six representatives of the rose family, *Dryas integrifolia*, *Sibbaldia procumbens*, *Potentilla nivea*, *P. emarginata*, *P. maculata*, and *P. tridentata*. The geographic range of the last reaches south to Cape Cod, the Alleghany Mountains, and northern Iowa.

Dr. E. K. Kane, in the two Grinnell expeditions—the first in 1850 and the second in 1853–'55—collected 106 species of flowering plants along the coast of Greenland from Sukkertoppen to the latitude of 81°. These plants are preserved in the herbarium of the Philadelphia Academy of Natural Sciences. According to Durand's report of Dr. Kane's collections, 73 species were from the coast north of Upernivik and the parallel of 73°, while 85 species were collected at Upernivik and southward, 52 of these being also found in the region farther north. Giesecke's catalogue, published in Brewster's Edinburgh Encyclopædia in 1832, enumerated 171 phænogamous species of Greenland; and two years earlier E. Meyer, in his *Plantæ Labradoricæ*, noted 224 phænogams, of which the greater part are indigenous in both Labrador and Greenland. Mr. Durand's compilation in 1856, from these and all other available sources, gave a total of 264 phænogamous plants occurring in Greenland, belonging to 109 genera and 36 families; of which 76 species, in 44 genera and 20 families, were recorded north of latitude 73°. The cryptogams of Dr. Kane's collections, also reported by Durand, comprised 42 species, being one Equisetum, three ferns, three species of

Lycopodium, twenty-five mosses, four species in the class Hepaticæ, and six thallophytes (lichens).*

Dr. Isaac I. Hayes, in his polar voyage in 1860–'61, wintering at Port Foulke (latitude 78° 10'), with a sledge journey northward along the east coast of Grinnell Land to Lady Franklin Bay (latitude 81° 40'), from which the return was made late in May, collected several thousand botanical specimens in that region northward from Whale Sound and Inglefield Gulf (latitude 77° to 78°). His plants, mostly collected at Port Foulke, in Greenland, are referred by Durand to 52 species, representing 36 genera and 18 families. They include five or six species that were not in Dr. Kane's list, and about a dozen which that list had not recorded for the tract north of 73°. The orders or families most largely represented in this far northern region are the Cruciferæ (mustard or cress family), 4 genera, with 8 species; the Caryophyllaceæ (pink family), 5 genera, with 7 species; the Rosaceæ (rose family), 3 genera, 5 species; the saxifrage family, represented by 7 species of saxifrage; and the Gramineæ (grass family), having there 5 genera, with 6 species. It is quite notable that the very large genus Carex (sedge), which has about 150 species in the Northern United States east of the Mississippi, and which has 44 species in southern Greenland, is represented in this very thorough observation and collection of the flora north of Inglefield Gulf by only one species.†

A very complete synopsis of the Greenland flora is given by Dr. Henry Rink in his valuable work, to which reference has been already so frequently made in

---

* Appendix xviii of Kane's Arctic Explorations. 1856, pp. 442–467.

† The Open Polar Sea, I. I. Hayes, 1867, pp. 398, 399,

this chapter. When his book was first issued in Denmark in 1857, the enumeration of the phænogamous or flowering plants comprised 320 species; but in 1877, when his new edition, and its translation into English with the aid of the distinguished botanist Dr. Robert Brown, appeared, this list had been enlarged to 361. Eighteen genera of the grass family are represented, with a total of 44 species. The largest genera are Poa, with 10 species, and Glyceria, 7 species. Both these genera are still more largely represented in the northern United States. The total of the Cyperaceæ (sedge family) is 51 species, in 6 genera, one of which (Carex), as before noted, has 44 species. Of the rush family, the genus Juncus has 9 species, and Luzula 6 species. The cosmopolitan orchis family, though best developed in warm temperate and tropical countries, notably in the Himalaya Mountains, has 4 genera (5 species) in Greenland· Birches are represented by 4 species, and there are 6 species of willows. The Compositæ, including daisies and dandelions, have 22 Greenland species, in 10 genera, the Hieracium, or hawkweed, having there 6 species. The Scrophulariaceæ (figwort family) have 6 genera, one (Pedicularis) having 7 species. Among the 10 genera of the heath family, the most numerous is the Pyrola (wintergreen), with 4 species. The genus Saxifraga, the only one of its family, has 12 species. The Ranunculaceæ (buttercup or crowfoot family) have 5 genera, each of one species in Greenland, excepting Ranunculus, which has 10. In the cress family, 13 genera comprise 25 species, of which 9 belong to Draba (whitlow grass). There are 3 species of gentian, two primroses, a poppy (extending to northern Greenland and Grinnell Land), and 3 species of violets. Nine genera of the pink family have 27 species, the most numerous being Stellaria (starwort), 6 species; Sagina (pearlwort), 5; and Alsine and Cerastium, each having 4 species. Finally, the rose family is represented by 17 species, in 7 genera, of which Potentilla (cinquefoil or five-finger) claims 8 species, or nearly the same number as are found in New England.

These details may convey the impression that the land border of Greenland, outside its ice sheet, has a somewhat rich flora; but this is the reverse of the truth, whether it is compared with the flora of Canada and the Northern States or with other arctic regions, as northern Europe, or the northern part of our continent and the adjoining archipelago. In this place we will only note that the flora of New England has about 1,300 native phænogams or flowering plants; that Minnesota has about 1,600; and California about 2,500.

Of acrogenous cryptogams, or the ferns and their allies, Greenland has 24 species. Thirteen are ferns in 6 genera, Polypodium and Woodsia having each 3 species. The genus Equisetum (horsetail) is represented by 4 species, and Lycopodium (club moss or trailing evergreen) by 5 species.

In the lower orders of cryptogams, Dr. Robert Brown, in an appendix of his English edition of Rink's work, enumerates 231 mosses and liverworts, 90 marine and fresh-water algæ, 203 lichens, and 10 fungi, or a total of 534 species. The list of fungi, however, especially of moulds and blights parasitic on the plants of higher orders, may certainly be very greatly increased by further special collections and studies. Even on the dried plants brought by the German expedition to eastern Greenland a dozen such parasitic fungi were identified.

As one of the glaciers in the Alps has a "garden" inclosed by the divided ice stream, so the nunataks, or tops of hills and mountains projecting above the Green-

land ice sheet, have their summer greenery and flowers.
For example, on Jensen's nunataks, a cluster of rocky
peaks rising 100 to 500 feet above the inland ice, at a distance of nearly 50 miles back (east-northeastward) from
the foot of the Frederikshaab glacier (latitude 62° 30')
and 20 miles from the nearest land outside the ice sheet,
Kornerup, the geologist and botanist of Lieutenant Jensen's party, in 1878, collected 27 species of flowering
plants. The ice surface there is 4,900 to 5,150 feet, or
nearly one mile, above the sea; and the nunatak summits vary in height from 5,200 to 5,650 feet. This very
high and isolated flora comprised an abundance of *Luzula hyperborea* and *Carex nardina*; the grasses *Trisetum subspicatum* and *Poa trichopoda*, in scattered tufts;
the sorrel, *Oxyria digyna*; the white-flowering *Cerastium alpinum* and *Saxifraga oppositifolia*; the little
blue-flowering *Campanula uniflora*; *Potentilla nivea*,
Ranunculus pygmaeus, *Silene acaulis*, *Cassiope hyp-
noides*, and *Armeria Sibirica*; and the very hardy, yel-
low-flowering arctic poppy, *Papaver nudicaule*, was
growing on the top of the highest nunatak. The same
species of Oxyria, Trisetum, Silene, and Cassiope are
among the arctic plants left stranded at the close of the
Glacial period in New England on the White Mountains, while this most plentiful Greenland saxifrage is
similarly found on the Green Mountains. Their altitude
on the Greenland nunataks is closely the same as on
these mountains, 1,300 miles farther south.

Near latitude 71° 40', on the north side of a mountain ridge which rises to the altitude of about 5,000
feet in a distance of five miles, bordering the Umanak
Fiord, Dr. Rink relates that he found flowering plants
to the height of 4,500 to 4,700 feet above the sea. His
ascent was on July 30, 1851, during a cold and unpleas-

ant summer. Concerning the vegetation of this mountain slope, and the limits of the perpetual snow and ice, he writes as follows:

The foreland consisted of low rocky hills alternating with fresh green meadowlike glens, and exhibiting the usual shrubs, such as the willow, crowberry, and Andromeda. Crossing a plain scattered over with huge boulders, we arrive at a somewhat steeper slope or terrace, on the top of which, at a height of 1,000 feet, the clouds are often seen to lie, enveloping the upper part of the mountains. It generally happens that when the sky has become clear after rains in September everything above this line appears scattered over with snow, while at the same time scarcely any snow has fallen on the lower land. Nevertheless, the surface was found to continue almost unaltered up to a height of 2,000 feet; the ground, consisting of gravel and clay, was covered by a thick sod containing almost the same plants as the low land. Only here and there a small heap of snow was concealed in some of the sheltered ravines; but at a height of between 2,000 and 3,000 feet the vegetable covering seemed decidedly thinner, the grasses and Cyperaceæ, which form the chief part of the sod, disappear, and are succeeded by mosses. At a height of 3,000 feet the same mosses still entirely cover small boggy places adorned with blooming buttercups. But on arriving at 4,000 feet the vegetation ceases to form any continuous mass, the plants standing singly in the gravel while the flat hollows are totally barren. The arctic willow finally disappears here, and several heaps of snow lie scattered over the ground, their under sheet consisting of solid ice. Footprints of reindeer and very old antlers were found here. Lastly, the increasing patches of snow at a height of 4,700 feet join to make a continuous sheet, leaving no ground visible. Close to the borders of the perpetual snowfield, among the numerous hillocks of ice and snow, at a height of more than 4,500 feet, the following plants were collected: *Papaver nudicaule, Potentilla Vahliana, Saxifraga tricuspidata, S. oppositifolia, S. cæspitosa, Alsine rubella, Silene acaulis, Draba arctica, Festuca brevifolia, Carex nardina.* Some scanty and stunted specimens of lichens, too defective to admit of their species being duly determined, were also found.

If we consider facts so apparently incompatible as plants in

bloom at a height of 4,500 feet in 71° north latitude, and snow re-
sisting the thawing of a whole summer's heat in 64 north latitude
at less than 100 feet above the level of the sea, it will appear a
rather difficult task to determine anything like a positive snow
line. In fact, the formation of glaciers from snow, owing to its
having lasted for several years and turned into solid ice, depends
on many different local influences besides the height above the
level of the sea. But still, on regarding the country as a whole,
a certain degree of uniformity may be discovered in the appearance of the more considerable mountain ridges. It will be found
that, at a height of from 2,000 to 2,200 feet or more, flat surfaces
of some extent and more or less excavated surfaces have in most
cases given rise to fields of perpetual snow and ice. From these
accumulations glaciers arise, and are moved onwards even to the
edge of the water, but glaciers originating at a lower height than
this are only exceptional. In the same way clear ground above
3,000 feet will be exceptional, and found to be caused by steepness, limited extent of a horizontal surface, and other accidental
influences.[*]

Even the surface of the snow and glaciers and of the
ice sheet itself has in many places its own very minute
cryptogamic plants. Some of these are found only on
the snow and *néré* fields, and one (*Protococcus nivalis*)
is the cause of " red snow," and, according to Chamberlin, is accountable, with red lichens, for the colour of the
Crimson Cliffs (latitude 76°) northwest of Cape York.
Six species of these snow algæ, as they may be called,
were collected by Nordenskjöld and Berggren during
their journey, in July, 1870, upon the inland ice near
latitude 68° 20'. Two of these species, however, seemed
to live only where the dust blown from the land of the
west coast (but supposed by Nordenskjöld to be cosmic
dust and named by him " kryoconite ") was collected in
very plentiful little pits of the ice surface.

---

[*] Danish Greenland, pp. 65-67.

Discussing the areal distribution of arctic phænogamous plants, Dr. Joseph D. Hooker found their whole number of species known to grow within the arctic circle to be about 770. Among them all he notes that the *Saxifraga oppositifolia*, so common in Greenland, "is probably the most ubiquitous, and may be considered the commonest and most arctic flowering plant." Only eight or nine species are peculiar to the arctic zone, and the remaining 762 (214 monocotyledons and 548 dicotyledons) extend also south of this zone. About 600 advance beyond latitude 40° north in some parts of the world, and about 50 of these are identified as natives of the mountainous regions of the tropics, while 105 occur in the south temperate zone.

The explored circumpolar area of the arctic flora has a width of 10° to 14° in latitude. "The only abrupt change," writes Professor Asa Gray, "in the vegetation anywhere along this belt is at Baffin Bay, the opposite shores of which present . . . an almost purely European flora on the east coast, but a large admixture of purely American species on the west." The portion of the whole belt richest in its diversity of vegetation is northern Scandinavia, or Lapland, which, though a small tract, has, according to Hooker, three fourths of the entire number of species and almost all the genera of this arctic area. He further remarks, indeed, that the Scandinavian flora not only girdles the globe above the arctic circle, and is predominant throughout the north temperate zone of the Old World, but also intrudes conspicuously into every other temperate flora, whether in the northern or southern hemisphere or on the mountains of tropical countries. This migration has been most favoured in America, where the Andes-Cordilleran mountain belt has permitted a considerable number of arctic and northern plants to extend also to the highest southern latitudes.

Greenland is especially cited by Hooker as having a flora closely identical, so far as it goes, with that of Lapland, but the latter is far richer in species. All but eleven of the 207 species of flowering plants (67 monocotyledons and 140 dicotyledons) recorded in Greenland north of the arctic circle, and likewise nearly all the species of its more southern part, are also native in northern Europe. Of the eleven not found there, three are known elsewhere only in Asia, and the remainder are North American, chiefly restricted to Labrador or to mountains farther south. Fifty-seven arctic plants of Greenland are absent, so far as shown by collections, from the same latitudes in North America and in the archipelago west of Baffin Bay. The eastern portion of the arctic area of North America, however, has 165 species which are not found in Greenland. The meridian of Davis Strait and Baffin Bay, therefore, appears to mark the limits of the migration of the Scandinavian flora outward both to the east, through northern Asia and North America, and to the west, by way of the Färöe Islands and Iceland, to Greenland. We can not doubt that the migrations took place upon continuous land areas, and that the contrast on the opposite sides of Baffin Bay has been caused by the long existence of a barrier of water there nearly as now.

How a land bridge was provided across the north Atlantic area and across the site of the present narrow and shallow Behring Strait will be considered in a later chapter; but we may here profitably notice, although briefly, the enforced migrations of arctic and boreal plants caused by the accumulation of ice sheets on the

northern lands during the Glacial period. Hooker concludes that "the existing Scandinavian flora is of great
antiquity, and that previous to the glacial epoch it was
more uniformly distributed over the Polar Zone than it
is now; secondly, that during the advent of the glacial
period the Scandinavian vegetation was driven southward in every longitude, and even across the tropics
into the south temperate zone; and that on the succeeding warmth of the present epoch, those species that survived both ascended the mountains of the warmer zones
and also returned northward, accompanied by aborigines
of the countries they had invaded during their southern
migration."

This view is shown by Hooker and Gray to supply
a satisfactory explanation of the almost complete identity of the Greenland flora with that of Lapland; its
less number of species, and its poverty of forms not also
found elsewhere; the rarity of distinctively American
species there; the fewness of temperate plants in the
somewhat temperate southern part of Greenland; and
the presence of a few of the rarest Greenland and Scandinavian species in very remote alpine localities of New
England and the Rocky Mountain region.   Dr. Hooker
reasons thus:

> If it be granted that the polar area was once occupied by the
> Scandinavian flora, and that the cold of the glacial epoch did
> drive this vegetation southwards, it is evident that the Greenland
> individuals, from being confined to a peninsula, would have been
> exposed to very different conditions from those of the great con-
> tinents.   In Greenland many species would, as it were, be driven
> into the sea, that is, exterminated; and the survivors would be
> confined to the southern portion of the peninsula, and, not being
> there brought into competition with other types, there could be
> no struggle for life amongst their progeny, and, consequently, no
> selection of better adapted varieties.   On the return of heat, sur-

vivors would simply travel northward, unaccompanied by the plants of any other country.*

In the first volume of the final reports of the Geological Survey of New Hampshire, Mr. William F. Flint and Dr. Nathan Barrows write of the arctic plants which occur on the summits of the White Mountains, but not elsewhere in that State and all the surrounding region. Dr. Barrows enumerates 52 flowering plants which are found in New Hampshire only on the White Mountains, and 57 others which likewise occur on the mountain tops above the limits of trees, but have also a more extended range on the lowlands. Forty of the 52 strictly alpine species are indigenous in northern Europe, and most of these occur also in Greenland. They were driven southward when the ice sheet was accumulated on the northern half of North America, and on the return of a warm climate, when the ice was melted away, were able to survive in the temperate latitudes only by finding congenial homes on the bleak, wind-swept, and cold mountain tops. The proportion (four fifths) of these mountain plants which are also European is far greater than for the whole flora of New England, in which only about one fifth part are also European species.†

The flora of Greenland is poor in its number of

---

* Outlines of the Distribution of Arctic Plants. Trans. Linnæan Society, London, vol. xxxii. 1861, pp. 251–348; reviewed by Prof. Asa Gray in the American Journal of Science, II, vol. xxxiv, pp. 144–148, July, 1862 (also in Scientific Papers of Asa Gray, 1889, vol. i. pp. 122–130).

† Besides the Geology of New Hampshire, vol. i, 1874, chapters xiii and xvii, see an article by Prof. J. H. Huntington, The Flowering Plants of the White Mountains, in Appalachia, vol. i, pp. 100–107, March, 1877.

species in comparison with the arctic part of America, and more so as compared with Iceland and Lapland. Hooker gives a list of 230 American boreal and arctic plants, none of which occur in Greenland, but of which 56 are found in Iceland, 57 in Europe, and 32 in the antarctic regions. Of the Iceland flora, 120 species are absent from Greenland, but 50 of these are found in northern Europe. Commenting on these remarkable features in the areal distribution of the circumpolar flora, Prof. Gray regards it as " most probable that the diffusion of species from the Old World to the New was eastward through Asia, for the arctic no less than (as has elsewhere been shown) for the temperate plants."

Subsequent to the studies of Hooker and Gray further important investigations of the Greenland flora and its history have been made by two Scandinavian botanists, Lange and Warming. In 1880 Lange tabulated 386 species of Greenland plants, of which he shows that 15 are endemic, that is, restricted to Greenland; that 40 are distinctively western or American but absent from Europe; and that 44 belong similarly to the eastern or European district, not occurring west of Baffin Bay. According to the researches of Warming, in 1889, Greenland may be divided, for classification of its flora, into two botanical provinces: a southern region characterized by the presence of the white birch, extending from Cape Farewell about two degrees, or 140 miles northward; and a second, more decidedly arctic region, comprising all the rest of the country. The small southern district has about 60 species of plants which are not otherwise found in Greenland, and many of these are especially European types. The larger region, or far the greater part of Greenland, has hardly any such Eu-

ropean plants, but a considerable number which are
peculiarly American. A large majority, however, have
a completely circumpolar range.

It is well argued by Warming that the mountains of
the western coast (7,000 feet high near the Umanak Fiord,
at latitude 71°, and 5,000 to 7,000 feet high near the
coast between latitude 62° and Cape Farewell, according
to Steenstrup and Holm, while farther inland there they
estimated the heights as probably 8,000 to 10,000 feet),
which rise above the limits that were reached by the
more extended glaciation during the Ice Age, and the
high Payer and Petermann peaks (7,000 and 11,000 feet
above the sea) near Franz Josef Fiord (latitude 73° 15')
in eastern Greenland, would permit the preglacial flora
to survive, at least partially, through the Glacial period,
so that it would spread afterward over the present land
borders. This view supplements that presented by
Hooker in ascribing the preservation of the flora partly
to *nunataks*, as well as to southward and then northward
migration on account of the formerly greater extent of
the ice sheet. While many plants in these ways survived the vicissitudes of the increased Pleistocene glaciation of Greenland, others succumbed, as the genera
Chrysosplenium, Caltha, Oxytropis, Astragalus, Phaca,
etc., which are so widely spread elsewhere in the arctic
regions and in alpine districts of temperate latitudes.*

---

* These notes of Lange and Warming are derived from Sir
Henry H. Howorth's review in the Geological Magazine, III, vol.
x, pp. 496–498, November, 1893. Howorth also adds that a similar conclusion is also reached by Nathorst in regard to Spitzbergen, whose present flora, instead of being due to immigration
since the Glacial period, is shown to be "the wreck and ruin of
what was once a much richer flora, and which has been able to
survive the drastic conditions which now prevail there."

Immediately before the Glacial period the flora of all Greenland was probably more allied with Europe than with North America, and this flora, excepting the very hardy species which could exist on nunataks, appears to have been largely preserved along the seashores. Greenland, like the northern half of North America, and like northern Europe, is known to have been much uplifted before the Ice Age, the high land elevation being doubtless the cause of the great accumulation of snow and ice. But then, as now, the Greenland ice sheet mainly lacked somewhat of reaching to the extreme shore line of the increased land area. Hence much of the preglacial flora which had spread westward from Scandinavia to the sea limit in Davis Strait and Baffin Bay escaped on the low coast (the most species being thus saved toward the south) and covered the open valleys again with verdure when the ice retreated. The broad glaciers, and even parts of the margin of the inland ice, extending quite to the sea in some places on both the east and west coasts, prevented the far northern migration of these plants; and in northern Greenland, besides such as could live there through the Glacial period, others have undoubtedly since come in, across the comparatively narrow channels, from the arctic archipelago and the American mainland.

More remotely, the lineage and history of the Greenland flora is revealed by the wonderfully abundant fossil plants which are found, with deposits of coal, in the rock strata of Disco Island and of contiguous and more northern islands and parts of the main western coasts between the latitudes of 69° and 73°. The plant-bearing beds are sandstone and shales, ranging in thickness up to 2,000 or 2,500 feet, and in age through the earlier, middle, and later parts of the Cretaceous period, to the

15

Miocene or middle period of the Tertiary era. Alternating with the upper beds, and covering them, are great outflows of columnar and amygdaloidal basalt, nearly horizontal, and varying in thickness up to 100 feet. The date of these lava outflows was approximately the same with similar or even grander volcanic action in the Färöe Islands, Iceland, and the region of the Cascade Mountains in Oregon and Washington; and in Iceland and the northwestern United States the volcanism has continued to recent times.

Prof. Lester F. Ward, reviewing the geographical distribution of fossil plants, in the Eighth Annual Report of the United States Geological Survey (for 1886–'87), summarizes the known fossil floras of Greenland, chiefly studied and figured by Heer, as follows: Carboniferous species, 1; Cretaceous species, 335, of which 88 belong to the Lower, 177 to the Middle, and 118 to the Upper Cretaceous, with 38 overlapping; and Tertiary species, 282,—giving a total of 618 fossil plants. The richest locality of the large Middle Cretaceous flora is Atanekerdluk (latitude 70°), the collections being obtained at the base of the high northeastern side of the Waigat passage, which separates Disco from the Nugsuak peninsula. Twelve hundred feet higher, directly up the same mountain side, are the richest Tertiary plant beds. For the Lower Cretaceous flora, one of the most productive localities is Kome (latitude 70° 37′), on the opposite northern side of the mountainous Nugsuak peninsula, whence Heer in 1871 described leaves of a poplar (*Populus primæva*), which, according to Ward, is "the most ancient dicotyledonous plant thus far published." About half of the Lower Cretaceous species are ferns, a quarter part are conifers, and a considerable number of the others are cycads, the one poplar being the only representative of our present predominant class of dicotyledons.

About two years after this review, extensive collections of the flora of the Potomac formation in Virginia and Maryland were described by Prof. W. M. Fontaine.* This formation, which is regarded as of Lower Cretaceous age, being perhaps as old as the Kome beds in Greenland, is very remarkable in containing a considerable proportion of dicotyledonous plants, the total published Potomac flora being 3 species of horsetail (Equisetum); 139 ferns; 148 gymnospermous plants, mostly conifers; and 75 dicotyledons. It still, however, seems not improbable that the primeval poplar of Kome is the earliest known forerunner of its great class.

From undetermined regions where these broad-leaved flowering plants had become well developed during earlier geologic periods, they suddenly appear on the scene in the Middle and Upper Cretaceous epochs. In the Middle Cretaceous flora of Atane, Greenland, fully half of the species are dicotyledons, including trees of several still living genera, as the poplar, fig, sassafras, persimmon, tulip tree, magnolia, and sumach. The somewhat later Upper Cretaceous beds of Patoot, Greenland, comprise 21 cryptogamous plants, all but two of which are ferns; 11 conifers, five of which belong to the genus Sequoia, a very widespread Cretaceous and Tertiary type, but now represented by only two living species, the redwoods or "big trees" of California; 5 monocotyledons; and 75 dicotyledons, or about five eighths of the whole number, including species of oak, walnut, plane tree or sycamore, laurel, cinnamon, aralia, dog-

---

* United States Geological Survey, Monograph XV, 1889.

wood, eucalyptus, ilex, buckthorn, cassia, and the other
genera of this class previously noted.

So late as the Miocene period, according to Heer's
correlations with the Tertiary floras of Europe, or, according to Sir William Dawson, J. S. Gardner, and
others, in the preceding Eocene period, the country
which is now the bleak and treeless Disco Island and
adjoining coast bore luxuriant forests chiefly composed
of plane trees and sequoias. Nearly half of the fossil
flora in Greenland referred by Heer to Miocene times
consists of trees, including 30 species of conifers,
besides many of our common deciduous broad-leaved
genera, as beeches, oaks, walnuts, poplars, maples, lindens, magnolias, and the plentiful plane trees, of which
last the stately sycamore, or buttonwood, in the eastern
half of the United States is a lonely surviving species
from the many of Tertiary times.

Not only is such a rich forest flora found to have
lived then within the arctic circle in Greenland, but
other leaf-bearing Tertiary deposits, and even a bed of
very good lignite coal 25 to 30 feet thick, were found
in 1876 by Capt. H. W. Feilden, the naturalist of the
English expedition commanded by Sir George Nares,
at latitude 81° 45', on the northeastern coast of Grinnell Land adjoining Robeson Channel. Among the
30 species of plants collected at that place and determined by Heer are three species of pine; two of spruce;
the bald cypress, very abundant, nearly like the same
species now living in the swamps of our Southern
States; two species each of poplar, birch, and hazel; an
elm; a viburnum; and a water lily. "There appear,"
Professor Heer writes, "in this most northern portion
of the earth, for the most part, the same species with
which we are already acquainted from Spitzbergen and

Greenland ; and it is highly probable that the same flora extended up to the Pole, and that, supposing dry land to have existed there, this latter was clothed with the same forest of coniferous and leafy trees." *

In Spitzbergen, Iceland, Greenland, and Grinnell Land, in British North America on the Mackenzie River at latitude 65° north, in Alaska, and in the New Siberia Islands, the fossil arctic Miocene flora, on which Heer did so much thorough study crowned with elaborate publication, has been found abundantly preserved. The latest accession to our knowledge of this flora was brought by Baron Toll, in 1886, from Thaddeus Island of the New Siberia or Lyakhoff group, where he discovered layers of Tertiary lignite and remains of Sequoia species.† Upon the arctic circle and much farther north, as known by the fossil plants in Grinnell Land, less than 600 miles from the pole, a temperate and even warm climate prevailed, nourishing forests in the now frigid zone on all sides of the pole during a geologic period not far preceding the Ice Age.

More free circulation of ocean currents than now, carrying the warmth of tropical regions into the circumpolar area, was probably the cause of the warm Tertiary climate. Ensuing uplifts of northern lands, to be discussed in another chapter, at length excluded the marine currents, and the increased land altitude, at least as great as the depth of the fiords, was attended with snowfall, instead of rains, till the snow and ice accumulation culminated in the Glacial period. During Tertiary times, according to Professor Asa Gray, " Greenland,

---

* Quarterly Journal of the Geological Society, London, vol. xxxiv. 1878, pp. 66–70.

† Nature, vol. xxxvii, p. 522, March 29, 1888.

Spitzbergen, and our arctic seashore had the climate of
Pennsylvania and Virginia now." Concerning the effect
of glaciation to disperse the luxuriant circumpolar Tertiary flora and its descendants southward, Professor Gray
further writes :

Here, then, we have reached a fair answer to the question how
the same or similar species of our trees came to be so dispersed
over such widely separated continents. The lands all diverge
from a polar centre, and their proximate portions—however different from their present configuration and extent, and however
changed at different times—were once the home of those trees,
where they flourished in a temperate climate. The cold period
which followed, and which doubtless came on by very slow degrees during ages of time, must have long before its culmination
brought down to our latitude, with the similar climate, the forest
they possess now, or rather the ancestors of it. . . . Wherefore
the high, and not the low, latitudes must be assumed as the birth-
place of our present flora, and the present arctic vegetation is best
regarded as a derivative of the temperate. This flora, which
when circumpolar was as nearly homogeneous round the high
latitudes as the arctic vegetation is now, when slowly translated
into lower latitudes would preserve its homogeneousness enough
to account for the actual distribution of the same and similar species round the world, and for the original endowment of Europe
with what we now call American types. It would also vary or be
selected from by the increasing differentiation of climate in the divergent continents, and on their different sides, in a way which
might well account for the present diversification.*

Another botanist, Dr. Leo Lesquereux, after extensive studies of American and foreign fossil floras, similarly concludes that "the essential types of our actual
flora are marked in the Cretaceous period, and have
come to us after passing, without notable changes,
through the Tertiary formations of our continent."

* Scientific Papers of Asa Gray, 1889, vol. ii, pp. 228, 229.

During long geologic ages, up to the Glacial period, Greenland possessed a mild temperate climate, and its shores bore as richly diversified a flora of forest trees and herbaceous plants as the present Northern United States. With the oncoming of the Ice Age, which still lingers in this bleak, mountainous, far northern land, most of its preglacial plant species, including all its large forest trees, perished, having no land pathway for escape, as from other parts of the circumpolar region, to the still temperate and even hot tropical and equatorial latitudes. After the partial mitigation of the severity of glaciation, as it continues within somewhat diminished limits in Greenland to-day, the plants which had survived on its low shores and on nunatak peaks again extended their range over all its land borders. Above the rock strata of Kome, Atane, Patoot, and the many other localities where the Cretaceous and Tertiary ancestors of the forests of the north temperate zone are found represented by their fossil leaves, the impoverished and dwarfed present Greenland flora blooms cheerily during its short summers, undaunted, and indeed made hardy and brave by its adversity, for patient endurance of its still glacial climate.

CHAPTER IX.

NATURALISTS have distinguished large regions of the earth which are characterized by peculiarities of their faunas and floras, certain species, genera, and orders, both of animals and plants, being limited in their geographic range so that they are plentiful in one region but absent or very scantily represented elsewhere. From such restrictions in zoölogical range, Sclater, in 1857, proposed six grand divisions of the earth's land areas. These great zoölogical regions, blending with each other on their boundaries, are now accepted by Wallace, Flower, and Lydekker, and many other writers on natural history; and each region is subdivided, on account of its climatic and topographic limitations of faunal and floral range, into smaller provinces. Sclater's divisions are: 1, The Palæarctic region, or the northern part of the Old World, comprising Europe, Iceland, Africa north of Sahara, and Asia north of the Himalayas; 2, the Ethiopian region, comprising intertropical and southern Africa, southern Arabia, and Madagascar; 3, the Oriental region, comprising India and extending southeastward to Java, Borneo, and the Philippine Islands; 4, the Australian region, including Celebes, New Guinea, Australia, New Zealand, and the islands of the Pacific Ocean; 5, the Nearctic region, or north part of the New World, including Greenland and

214

North America as far south as northern Mexico; and, 6, the Neotropical region, embracing the remainder of the American continent and the West Indies.

Other authors, as Heilprin and Packard, note so close relationship and community of species through all the far northern countries of both the Old and New Worlds, of which we have reviewed the botanical side in the foregoing pages, following Hooker and Gray, that they prefer to unite these northern areas of the Eastern and Western hemispheres as a single circumpolar zoölogical and botanical region, which is named the Holarctic by Heilprin, from its reaching wholly around the earth in the arctic zone.

From the geographic distribution of animals, not less than of plants, abundant evidence is found that in a late geologic time, probably comprising the closing stage of the Tertiary era and the early part of the Quaternary until the Ice Age, an extensive land area occupied the present place of Behring Strait and Sea, upon which the fauna and flora of the northern lands freely migrated from Europe and Asia to America, and the reverse, becoming nearly alike in these two great continental regions. Over all the circumpolar land expanse the mammoth, mastodon, and many other large animals roamed from the United States to Alaska, Siberia, Continental Europe, and the British Isles during late Tertiary times. Near the end of the latest Tertiary period, or more probably well forward in the Quaternary era, almost to the epoch when the increasing uplift of northern countries brought on the Ice Age, men, having been created through evolution from the anthropoid apes, spread outward from their native tropical portion of the Old World to all parts of the great land areas of that hemisphere and to America. On account of the

opportunity afforded in the circumpolar region, under
its mild Tertiary and early Quaternary climate, for free
intermigration of plants and animals, the present fauna
of Greenland, like its flora, has close alliance or identity
of many species with those of arctic Europe and
Siberia.

Because of the greater freedom of animals than of
plants to extend their geographic range within recent
times from the northern part of America and from the
arctic archipelago to Greenland, they give less decisive
evidence than the flora for the late Tertiary and preglacial union of Greenland with Iceland, the Färöe
Islands, Britain, and Europe, while the Baffin Bay
marine barrier prevented communication between Greenland and North America. As in the case of the plants,
many species of the preglacial land fauna in Greenland
probably survived on the shores and on the *nunataks*
during the more extended glaciation, which appears to
have been contemporaneous with the North American
and European Ice Age. Since that time other species
have doubtless come in, relatively in greater proportion
than for the flora, from the contiguous lands on the
west.

Only seven or eight species of land mammals are
known to exist in Greenland in a native or wild condition, to which the domestic cat and dog (perhaps to be
considered the same species as the domesticated Eskimo
dogs), goat, sheep, ox, hog, and house mouse, brought
into the country by the Danes, are to be added for the
total mammalian fauna.

A single specimen of the arctic wolf, probably astray
from the opposite western shore of Baffin Bay and
Smith Sound, was shot in 1869. The whole number of
Eskimo dogs in Greenland was estimated by Rink in

1877 to be about two thousand. He thinks them to have been derived, through domestication, from the nearly related arctic wolf; and Peary, after much experience in dog sledging across the inland ice, shares the same opinion.

The most important fur-bearing land animal is the arctic fox (*Canis lagopus*, genus *Leucocyon*, Gray), which is rather common, occurring in two varieties, called blue and white. They are hunted by the Eskimos for selling their skins to the Danish traders. The price of the white is small, but that of the rarer blue variety, when in its best pelage, has several times risen to fifteen dollars in the European market. Flower and Lydekker describe this species as having " the tail very full and bushy, and the soles of the feet densely furred below. Its colour changes, according to season, from bluish gray to pure white." Dr. Rink says:

The foxes appear mostly confined to the mainland, though they live, perhaps, for the most part upon what they may find on the shore at low water. In summer they often visit the islands, and may be met with, having made their holes and bred their young, in the immediate vicinity of abandoned winter houses, apparently attracted thither by the garbage left by the inhabitants. The half-grown cubs may then be seen playing outside like whelps of tame dogs, and may be approached and taken by the hand. They are always easily tamed, and behave in this state just like dogs. The skins are only in season from November to March, the hair being shorter and of a dirty grayish-brown colour during the other months. Many of the foxes are caught in traps of a very primitive construction, formed of a flat stone so fixed as to fall down and crush or confine them when the bait has been touched. But most of them are shot by hunters lying in wait when they come down to the shore in search of mussels or other food at low water. Of course, this sort of sport is limited to those southern places where the sea off the shore is mostly open even during the first part of the winter. There the foxes are by far the most numerous, but even in the farthest north they are not wanting. Wherever seals have been caught in winter upon the ice, footprints of some fox that has been attracted by the drops of blood in the snow will generally be visible; but how these animals are able to find the food necessary for supporting life during eight months of the year at the northern fiords remains somewhat of a mystery. The sea being frozen over for hundreds of miles, the beach especially being covered by a crust of ice upwards of ten feet in thickness, and the birds having migrated to the south, the only other animals left to roam over the vast snow-covered tracts, besides reindeer, are hares, partridges, ravens, eagles, hawks, and owls. All these terrestrial animals are scarce, and seem to be unable to yield sufficient food for the foxes. Severe winters with much snow are favourable to fox hunting, and such favourable years seem to have been often succeeded by periods in which the animals were unusually scarce. The number of foxes killed have been fifteen hundred yearly, on an average, from 1853 to 1872; the greatest number ever obtained seems to have been in 1874, when they amounted to five thousand. Of the whole stock, about one half is caught between 60 and 61 north latitude.*

The white or polar bear (*Ursus maritimus*), called the "water bear" in Labrador (from its life on the floe ice and often swimming in the sea), has a comparatively small head, with small and narrow molar teeth, and the soles of its feet are more covered with hair than in other species of its genus. It is less ferocious than has been generally supposed, and Hayes affirms that it has "never been known to attack man except when hotly pursued and driven to close quarters." Packard shows that these bears lived in the time of Cabot and Cartier, nearly four hundred years ago, as far southward as Newfoundland; but their numbers and geographic range are now much diminished. Dr. Rink's notes of this most characteristic arctic mammal, as observed in Greenland, are as follow :

---

* Danish Greenland, 1877, pp. 104, 105.

The polar bear is almost an amphibious animal. Upwards of fifty of them are, on an average, shot yearly, of which more than one half are shot in the environs of the northernmost settlement, and of the remainder the greater part at the southernmost extremity of the country, where they arrive with the drift ice around the Cape Farewell. Throughout the whole intervening tract bears are scarce, but still they may be found everywhere, and solitary stragglers may even be met with unexpectedly in summer in the interior of the fiords. Killing a bear has, in ancient as well as in modern times, been considered one of the most distinguishing feats of sportsmanship in Greenland. Erik the Red is said to have quarrelled with one of his best friends from envy on account of the latter having had the luck to capture a bear. . . . In the north the bear is pursued upon the frozen sea by the aid of dogs. It often takes refuge on the top of an iceberg, where it is surrounded and held at bay by the dogs until it is shot, generally not without some of the latter being lost on the occasion. In the north the male bears at least seem to roam about in winter as far south as 68 north latitude, because wherever the carcass of a whale may be found, or a rich hunt of seals or white whales occurs in a certain place within these confines, there several bears are sure soon to make their appearance. In the south, where no dogs are to be had for assistance, the natives generally try to force the bear into the water, and often kill it with harpoons from the kayak.*

The ermine (*Putorius ermineus*), the North American lemming (*Myodes obensis*), and the musk ox (*Ovibos moschatus*), "are not found," according to Dr. Robert Brown, "as far south as the Danish possessions. They inhabit the shores of Smith Sound and East Greenland in about the same latitude, but do not stretch farther south, so that the probabilities are that they have migrated round the northern end of the country, and are kept from spreading southward by the glaciers."

---

* *Op. cit.,* pp. 106, 107.

Arctic hares (*Lepus glacialis*) are rather infrequent; no more than a thousand, according to Rink's estimate, being killed by the Eskimos yearly.

Above all the other land animals in Greenland, as in all the arctic zone, the reindeer (*Rangifer tarandus*), extensively domesticated in northern Europe and Siberia, is the most interesting and useful to man. Flower and Lydekker say of this species :

> The reindeer, or caribou, as it is termed in North America, is the sole representative of the genus *Rangifer*, which is sufficiently distinguished from all its allies by the presence of antlers in both sexes. . . . This animal is distributed over the northern parts of Europe, Asia, and America, the differences which may be observable in specimens from different regions not being sufficient to allow of specific distinction. The reindeer is a heavily built animal, with short limbs, in which the lateral hoofs are well developed, and the cleft between the two main hoofs is very deep, so that these hoofs spread out as the animal traverses the snow-clad regions in which it dwells. The antlers are of very large relative size. There is a bez as well as a brow tine, which are peculiar in being branched or palmated. In the American race (caribou), as well as in some of the specimens found fossil in the English Pleistocene, one of the brow tines is generally aborted to allow of the great development of the other.*

In Greenland the reindeer has a geographic range from Foulke Fiord (latitude 78° 18'), where they were seen in small numbers by Hayes, and from the neighbouring northern shores of Inglefield Gulf, where Peary found considerable herds, southward along all the west coast to Cape Farewell. Rink gives the following description of reindeer hunting by the Eskimos in summer :

---

* An Introduction to the Study of Mammals, Living and Extinct, 1891, pp. 324, 325.

The chief hunting grounds, situated by the interior of the
fiords, were the rendezvous places for hunters from widely
spread wintering places. In their skin boats, escorted by kayaks,
they carried their families, tents, and all their necessaries. From
the fiords the boats occasionally were borne by land to the lakes,
and in this way the farthest accessible interior regions were
visited. . . . The chase gradually increased, chiefly on account of
the more common use of the rifle. It reached its culminating
point in the period between 1845 and 1849, when the number of
deer killed might be rated at 25,000 annually, the number of skins
exported being about 16,000 per annum. But after 1850 the chase
rapidly declined, and in 1868 to 1872 the annual export had
dwindled to 6 skins, or, in other words, nothing, while the whole
number of animals killed may scarcely be rated at more than
1,000 per annum during the latest period. On these longer excursions the hunters are generally accompanied by women, whose
duty is to carry home as much of the venison as possible, which
they effect by passing the strap from which the load hangs
around their foreheads. Notwithstanding the great skill they
have acquired in carrying burdens in this way, their assistance
proved insufficient during the briskest period of the chase. We
may suppose that during those years one half of the flesh was
abandoned on the rocks, while a great many deer were killed only
for the sake of the hide and the tongue. . . . We have before us
but an instance of a similar destruction of various kinds of game in
almost every other part of the globe. Now, since the animals have
ceased to congregate in so great herds, the pursuit of them never
can be so ruinous, and it seems that during the last years the remaining stock has not been subject to any further decrease. Reindeer meat has, of course, ceased to be in daily use, but, on the
other hand, it has as yet by no means become a rarity in Greenland. . . . The tallow is highly appreciated, and always eaten raw
as a titbit. The antlers are almost indispensable for the manufacture of several weapons and kayak implements. In the years
following 1850 they were made an article of trade and purchased
at a price of about one penny per pound, and the quantity of
them dispersed over the country proved to be so great that at one
station more than 100,000 pounds were brought for sale within a
short period. Although hunting the reindeer was ordinarily a
summer occupation, it also took place in winter, and in some

localities reindeer were even shot close to the houses. Exceptionally the inhabitants of some places made the chase by land their
chief source of sustenance, though not abandoning seal hunting,
to which they had recourse for the purpose of providing oil for
their lamps.*

Far more valuable to the Greenlander for the means
of subsistence in his frigid and barren, mountain-girt,
and mostly ice-enveloped country, are the aquatic mammals known to naturalists as the pinnipeds and the
cetaceans, the former comprising the walrus and seals,
while the latter include whales, dolphins, and porpoises.
Dr. Rink notes the uses of these animals by the Eskimos, and their methods of capture, as follows:

> These two different orders of mammalia . . . are caught in
> much the same manner, and . . . both their flesh and blubber
> supply the Greenlanders with their most nutritious food, and
> with the necessary means for heating and lighting their houses.
> But as regards their skins, there is an important difference, the
> seals affording material for clothes. boats, and tents, while the
> skin of the whales only yields a favourite nutriment. This edible
> skin is called Matak, and all the cetaceous animals are for this
> reason denominated Mataliks. It is almost always eaten raw, and
> consists of two sheets, the inner one very tough, while the outer,
> which is considered to represent the rudimental hair covering, is
> more brittle. There are as many varieties of it as there exist
> species of cetaceans, from that of the small porpoise to the colossal
> whale.
> The principle upon which the peculiar seal hunting of the
> Eskimo is based is almost the same as that which has been resorted
> to in the ordinary European mode of killing whales—viz., previ-
> ously striking the animal with the harpoon, and keeping hold of
> and wearying it by aid of a line attached to the harpoon, so as to
> facilitate the killing of it by means of the lance and spear. The
> use of the harpoons, lines, and lances is perfectly analogous among
> the European and the Eskimo hunting at sea. The chief differ-

---

* *Op. cit.*, pp. 101–104.

ence in their proceedings appears to be the mode of retarding the animal while running off with the harpoon and lines.

The Greenlander manages this by throwing out an inflated bladder attached to the other end of the line; but the European whaler still keeps this end of the line in the boat of the harpooner,

Fig. 17.—Kayak, with inflated bladder and other implements of chase.

only letting go so much of the line as is necessary to prevent the boat from being capsized and drawn down, while the terrified animal, being still in possession of its whole power, runs off with extraordinary quickness. The seal, or whale, having become sufficiently exhausted by dragging [the bladder or] the boat, the mortal wounds are finally inflicted by help of the lance.*

* *Op. cit.*, pp. 111, 112.

16

First in size and in uniqueness of specific characters among the pinnipeds is the walrus (*Odobænus ros-marus*, genus *Trichecus* of Gray and most writers during the past century), also called the morse, from its Russian and Lapp name, or the sea cow or sea horse. Its length is ten to twelve feet, and its weight two thousand to three thousand pounds. From the first discovery of the Atlantic shores of North America up to about a hundred years ago immense herds of walruses or " sea oxen " (assembling " to the amount of seven or eight thousand ") lived about Sable Island, off the coast of Nova Scotia, on the Magdalen Islands and Anticosti in the Gulf of St. Lawrence, and around Newfoundland. A thousand years ago the walrus was abundant on the coasts of Finmark, the most northern part of Norway. Relentless slaughter for its oil, hide, and tusks, has since restricted the limits of this species in the Western Hemisphere to Labrador and the coasts of Hudson and Davis Straits, Baffin Bay, and adjoining waters, and in the Old World to " the islands and the icy seas to the northward of eastern Europe and the neighbouring portions of western Asia, where it rarely, if ever, now visits the shores of the Continent." * A closely similar species, the only other of the genus, inhabits the region of Behring Sea and Strait and the contiguous shores of the Arctic Ocean.

In Greenland, according to Dr. Robert Brown, the walrus "is found all the year round, but not south of Rifkol, in latitude 65°. In an inlet called Irsortok it collects in considerable numbers, to the terror of the natives who have to pass that way. . . . It has been

* J. A. Allen, History of North American Pinnipeds, 1880, p. 79.

found as far north as the Eskimos live or explorers have
gone. . . . It is not now found in such numbers as it
once was, and no reasonable man who sees the slaughter to which it is subject in Spitzbergen and elsewhere can doubt that its days are numbered.   It has
already become extinct in several places where it was
once common.   Its utter extinction is a foregone conclusion." *

Rink says: " *The walrus* is only rarely met with
along the coast, with the exception of the tracts between
66° and 68° north latitude, where it occurs pretty nu-
merously at times.   The daring task of entering into
contest with this animal from the kayak on the open sea
forms a regular sport to the natives of Kangamiut in 66°
north latitude.   The number yearly killed has not been
separately calculated on account of the skin being generally eaten along with the meat, and considered a very
delicate dish; but they can hardly exceed two hundred." †

Considerable herds of the walrus were found by
Kane, Hayes, and Peary in the region of Smith Sound
and Inglefield Gulf, and they describe very exciting
encounters with these powerful animals, when, some of
their number being wounded or killed, they sought to
avenge their losses by concerted attacks on their perse-
cutors.

The chief food of this huge animal consists of bi-
valved molluscs, notably *Mya truncata* and *Saxicava
rugosa*, which are dug out from their beds in the mud
and sand of the sea bottom by means of the animal's
tusks.   " It crushes and removes the shells by the aid of

---

* Proceedings of the Zoölogical Society, London, 1868, p. 433.
† *Op. cit.*, pp. 126, 127.

its grinding teeth and tongue, swallowing only the soft
part. . . . It also feeds on other molluscs, sandworms,
starfishes, and shrimps."

Among the five or six species of seals of Greenland,
the most abundant and important to the natives as a

Fig. 48.—Women dressing a ringed seal just brought in by a kayaker.

source of food is the *Phoca fœtida*, or ringed seal, of
naturalists, called the fiord seal by Fabricius, the *natsek*
of the Eskimos, and the "floe rat" of English whalers.
Its length is four to six feet, and its weight ranges from
seventy-five to two hundred pounds or more. Allen
says of this species:

The ringed seal is pre-eminently boreal, its home being almost exclusively the icy seas of the arctic regions. Its favourite resorts are said to be retired bays and fiords, in which it remains so long as they are filled with firm ice: when this breaks up they betake themselves to the floes, where they bring forth their young. It is essentially a littoral, or rather glacial species, being seldom met with in the open sea.*

Its geographic range extends south to Labrador, and in Europe to the Orkneys, the Hebrides, and the Gulfs of Bothnia and Finland. Eastward it ranges along the coasts of Siberia and Alaska, and extends south into Behring Sea. It is said to occur also in the wholly fresh water of Lake Ladoga in Russia; and near allies of this species live in the Caspian Sea, and in the fresh Lake Baikal, 1,360 feet above the level of the ocean.

Dr. Rink estimates the number of these seals captured yearly in Greenland to be about 51,000. He remarks on its habits and the origin of its specific name:

Stray individuals of this species migrate to the main drift ice of Baffin Bay in July, and return to the coast when the first bay ice is forming in September, or occasionally appearing whenever the weather has been stormy. But the chief stock, whose favourite haunts. as has been described, are the ice fiords, does not seem to leave the coast at all. It is almost exclusively this seal that is captured as "utok" and by means of the ice nets. It derives its scientific name from the nauseous smell peculiar to certain older individuals, especially those captured in the interior ice fiords, which are also on an average perhaps twice as large as those generally occurring off the outer shores. When brought into a hut and cut up on its floor, such a seal emits a smell resembling something between that of asafœtida and onions, almost insupportable to strangers. This peculiarity is not noticeable in the younger specimens or those of a smaller size, such as are more generally caught. and at all events the smell does not detract from the utility of the flesh over the whole of Greenland.†

---

* *Op. cit.*, p. 619. † *Op. cit.*, p. 123.

Quite inferior in its value to the Greenland Eskimos is the *Phoca vitulina*, or harbour seal, called in Greenland the *kassigiak*, or "spotted seal," and known commonly in Norway as the "fiord seal." Its length is nearly like the preceding, and its average weight is from sixty to one hundred pounds. Southward its geographic range extends to New Jersey, and stray specimens have been captured on the coast of North Carolina. On the European coasts it reaches south to the British Isles, Spain, and even the Mediterranean, and "along the Scandinavian peninsula is the commonest species of the family." Thence it extends along the northern shores of Europe and Asia, through Behring Strait, and southward to Japan and to the Santa Barbara Islands, off the coast of southern California. It also ascends the larger rivers of these coasts, often to a considerable distance above tide water, having been taken rarely in Lakes Champlain and Ontario. It is the only species of seal inhabiting any part of the eastern United States. Allen writes:

On the New England coast, as elsewhere, it is chiefly observed about rocky islands and shores, at the mouths of rivers, and in sheltered bays, where it is always an object of interest. Although ranging far into the arctic regions, it is everywhere said to be a sedentary or non-migratory species, being resident throughout the year at all points of its extended habitat. Unlike most of the other species, it is strictly confined to the shores, never resorting to the ice floes, and is consequently never met with far out at sea, nor does it habitually associate with other species.*

Dr. Rink states that this species is " much less numerous in Greenland than the natsek. It occurs here and there, however, throughout the coast, and seems to be as stationary as the former, with which it also corre-

* Page 589.

sponds in size. The skin is highly valued in Greenland for making clothes. The annual catch is doubtful; it may be guessed at a thousand, and at any rate scarcely amounts to two thousand." *

Next after the ringed seal, the most abundant and valuable species for the Eskimos is the *Phoca grœnlandica*, or harp seal, commonly called the "saddleback" by English-speaking whalers and the *svartside* by the Danes, but known to the Norwegians and Swedes as the Greenland seal. This gregarious and migratory species grows to be five or six feet long, with weight varying from three hundred to seven hundred pounds, of which the skin and blubber constitute about a third part. Allen writes of its geographic range and migrations:

Although the harp seal has a circumpolar distribution, it appears not to advance so far northward as the ringed seal or the bearded seal; yet the icy seas of the North are pre-eminently its home. It is not found on the Atlantic coast of North America in any numbers south of Newfoundland. A few are taken at the Magdalen Islands, and while on their way to the Grand Banks some must pass very near the Nova Scotia coast. . . . The harp seals are well known to be periodically exceedingly abundant along the shores of Newfoundland, where during spring hundreds of thousands are annually killed. In their migrations they pass along the coast of Labrador, and appear with regularity twice a year off the coast of southern Greenland. . . . The saddleback, although found at one season or another throughout a wide extent of the arctic seas, appears to be nowhere resident the whole year. Its very extended periodical migrations relate apparently to the selection of suitable conditions for the production of its young, and occur with great regularity. Where it spends portions of the year is not well known, while, on the other hand, it may be found with the utmost certainty at particular localities during the breeding season. Its most noted breeding stations are

---

* Pages 123, 124.

230 GREENLAND ICEFIELDS.

the ice floes to the eastward of Newfoundland and in the vicinity of Jan Mayen, at which localities they appear early in spring in immense herds.*

Notes of the harp seal, as observed in Greenland by Dr. Rink, are as follows:

This species, which is well known as forming the chief object of chase to the European sealing ships in the Spitzbergen and Newfoundland seas, is a migratory animal, but must nevertheless be considered at home on the Greenland coast, on account of its haunting its shores and roaming over its sounds and fiords regularly during the greater part of the year. It is of inestimable importance on account of its skin, which yields the usual covering of the kayaks and open skin boats. It appears regularly along the southern part of the coast in September, travelling in herds from south to north between the islands, and at times resorting to the fiords. They are then pretty fat, but their sheet of blubber is still increased during the course of winter. In October and November the catch is most plentiful; then it decreases in December, grows more scarce in January, and becomes almost extinct in February. The seals return as regularly off the southern coast in May, and on the more northern in June. They have then grown very lean and lost more than half their blubber. Having visited the fiords in numerous herds, they again disappear in July and return in September. Consequently this seal deserts the coast twice a year, and as regularly returns to it in due season, always first making its appearance in the southern and somewhat later in the northern regions. . . . The "saddlebacks," according to their age or different stages of development, are divided into four or five different classes by the European sealers, as well as by the natives of Greenland. In Greenland, however, in familiar language, the only distinction made is between the full-grown animal, whose skin has assumed the half-moon-shaped dark marking on both sides of the body, and the half-grown ones, or "blue-sides," on whom these bands are as yet not sufficiently developed. . . . The annual catch is calculated at 17,500 full-grown saddlebacks and 15,500 blue-sides.†

---

* Pages 640–642.   † Pages 124–126.

It seems very doubtful whether the gray seal (*Halichærus grypus*) has ever been found on the western coast of Greenland, but it is probably frequent on the east coast. It ranges from Nova Scotia, Labrador, and the British Isles northward to Iceland, and to Tromsö in Norway, but is absent from the islands of the Arctic Ocean.

The bearded seal (*Erignathus barbatus*), called the *ugsuk* or "thong seal" in Greenland, is, after the walrus, the largest species of our North Atlantic pinnipeds. Its length is stated by Rink as ten feet, and he notes that it "occurs only in few numbers, and chiefly at the northern and southern extremities of the coast, but is of the utmost importance, its big skin being the only one considered fit for making the hunting lines of the kayakers, whose life depends on the line running out easily without being liable to the slightest entanglement when pulled by the harpooned seal. The annual catch hardly amounts to a thousand." Allen says of its range: "The present species is circumpolar and extremely boreal in its distribution, and appears to be migratory only as it is forced southward in winter by the extension of the unbroken icefields."

The only other species of seal frequenting the shores of Greenland is the "bladder-nose," or hooded seal (*Cystophora cristata*), "well known from the bladder on its forehead, which it is able to blow up at will." Its length is six to eight feet, with weight of four hundred to nine hundred pounds, the skin and fat being nearly half. Allen says of this species: "The hooded or crested seal is restricted to the colder parts of the North Atlantic and to portions of the Arctic Sea. It ranges from Greenland eastward to Spitzbergen and along the arctic coast of Europe, but is rarely found

south of southern Norway and Newfoundland. . . . Like
the harp seal, it appears also to be regularly migratory,
but, owing to its much smaller numbers and less commercial importance, its movements are not so well
known."

Nansen describes the ruthless slaughter of this species on the floe ice between Iceland and Greenland as
follows :

The capture of the bladder-nose in Denmark Strait is not an
industry of very long standing. It was inaugurated by the Norwegians in 1876, and their example was followed by a few English
and American vessels. For the first eight years the venture was
an unprecedented success: the seals were more than plentiful,
and were shot down in thousands. During this period something
like five hundred thousand head were captured, and it is probable
that quite as many were killed and lost. After these years of
plenty came a change, and ever since the pursuit has been practically a failure, all the vessels alike being equally unsuccessful.*

In 1888, however, Nansen saw again vast numbers
of these seals on the inner and firm areas of the floe ice
in the same region, where, he thinks, they had learned
to stay as less exposed to their human enemies, instead
of living as formerly on the outer and loose floes, where
they better escaped the ravages of polar bears.

Notes of the hooded seal on the western coast of
Greenland, by Rink, are as follows :

It is only occasionally found along the greater part of the
coast, but visits the very limited tract between 60° and 61° north
latitude in great numbers, most probably coming from and returning to the east side of Greenland. The first time it visits us
is from about May 20 till the end of June, during which it yields
a very lucrative catch. It is very fierce, and when wounded not
unfrequently attacks its pursuer, violently splashing, and trying

---

* The First Crossing of Greenland, 1890, vol. i. p. 182.

to bite him. This hunt, which is hazardous to a man in a frail kayak, has been greatly facilitated by the rifle, the hunters first hitting their prey from the ice floes, and afterwards despatching it with their harpoon from the kayak. A bladder-nose yields about one hundred and twenty pounds of blubber and two hundred pounds of flesh. The annual catch is about three thousand on an average.*

Sixteen species of cetaceans are enumerated by Dr. Robert Brown as known to occur in the waters of western Greenland. The most valuable of these to the Eskimos is the white whale (*Beluga catodon*, or *Delphinapterus leucas*), which, according to Dr. Rink, swims along the coast "chiefly in spring, as soon as the bay ice breaks up, and in autumn before the new ice forms. It measures twelve to sixteen feet in length, and yields about four hundred pounds of blubber and an equal or greater amount of eatable parts. The number yearly killed may be estimated at more than six hundred." The geographic range of this small whale extends south to the Gulf of St. Lawrence, from which it ascends the River St. Lawrence for a considerable distance. During the Champlain epoch or closing part of the Glacial period, when the sea extended inland to the basin of Lake Champlain as soon as it was uncovered from the waning ice sheet, this species ranged along the then enlarged Gulf of St. Lawrence to Vermont, where a nearly complete skeleton of it was found in 1849, embedded in the Champlain marine clays, sixty feet above the lake, and nearly one hundred and sixty feet above the present sea-level. At several localities in Canada, also, along the St. Lawrence Valley, the bones of this species have been found in the Champlain clays, up to heights exceeding two hundred feet above the sea.

---

* *Op. cit.*, p. 126.

Second in value to the Greenlanders is the narwhal or " sea unicorn " (*Monodon monoceros*), so named from the long and straight, spirally grooved, single tusk or " horn " of the male, which often attains a length of seven or eight feet. It is " essentially an arctic animal, frequenting the icy circumpolar seas, and but rarely seen south of 65° north latitude." Rink tells us that it is much scarcer than the white whale, and is " almost only caught at the northernmost settlements, especially in Umanak Bay. It follows immediately after the white whale, and is chased from the kayak in November, when the surface of the sea every moment threatens to be rapidly congealed in calm weather to the utmost danger of the hunters. The annual catch probably does not surpass a hundred."

The large species of whales are now less frequently captured by the Eskimos than formerly, the present average number taken yearly being no more than two or three. These colossal marine mammals, the chief objects of pursuit by the European and American whalers, include the Greenland or arctic right whale (*Balæna mysticetus*), forty-five to fifty feet long; three species of rorquals, " finners," or " razorbacks," as they are variously called, namely, the " blue whale " (*Balænoptera sibbaldi*), " the largest of all known animals," having a length of about eighty feet, and two others of this genus, commonly called the " big finner " and the " little finner "; the humpback whale (*Megaptera longimana* or *Megaptera boops*), about fifty feet long; and the sperm whale (*Physeter macrocephalus*), very rare in Baffin Bay, of which the male grows to be sixty feet long, while the female is only about half as large. The sperm whale is mainly a tropical species; but the " humpback " is chiefly limited to far northern latitudes, as from Norway to Baffin Bay. During the closing stage of the Ice Age, however, the humpback whale is known to have lived in the Gulf of St. Lawrence, for portions of a skeleton of this species, as reported by Sir William Dawson, were found in 1882 in the gravel and sand of a ballast pit on the Canadian Pacific Railway, three miles north of Smith's Falls, in Ontario, and thirty miles north of the River St. Lawrence. This locality is four hundred and forty feet above the sea, having nearly the same height as one of the principal Late Glacial or Champlain marine shore lines on the Montreal Mountain and in other parts of the St. Lawrence Valley.* Even at the present time this species is occasionally seen in the Gulf of St. Lawrence, and, according to Dawson, "is more disposed than the other large whales to extend its excursions some distance into the estuary."

Among the other and comparatively small cetaceans of Greenland, besides the right whale and narwhal before noticed, are the bottle-nose whale (*Hyperoödon rostratus*); the pilot whale or *Grindhval* of the Färöe Islands (*Globicephalus melas*); the common porpoise (*Phocæna communis*), of which a few are usually obtained by the Eskimo hunters yearly; two or three others of the dolphin family, observed still more rarely; and the grampus or "killer" (*Orca gladiator*), "readily known, when swimming in the water, by the high, erect, falcate dorsal fin, whence their common German name of *Schwertfisch* (swordfish). . . . They are distinguished from all their allies by their great strength and ferocity, being the only cetaceans which habitually prey on warm-blooded animals, for, though fish form part of their food,

---

* J. W. Dawson, The Canadian Ice Age, 1893, p. 268.

they also attack and devour seals and various species of their own order, not only the smaller porpoises and dolphins, but even full-sized whales, which last they combine in packs to hunt down and destroy, as wolves do the larger ruminants." *

Long lists of the fauna of Greenland, in its other great classes, are given by the Manual of the Natural History, Geology, and Physics of Greenland and Neighbouring Regions, prepared in 1875, under the editorship of Prof. T. Rupert Jones, for the use of the British Arctic Expedition in 1875–'76, commanded by Sir George Nares; and these lists, with additions, are also given in an appendix of Dr. Henry Rink's Danish Greenland, as edited in English by Dr. Robert Brown in 1877.

The Greenland avifauna, according to the list by Reinhardt, thus published, comprises 124 species, of which 51 species are very rare, or occur only as stragglers far from their ordinary geographic range. Among the rare or astray species which are familiar summer birds in the eastern United States, are the golden-winged woodpecker or flicker, several flycatchers and warblers, and even the robin. In the grouse family, the rock ptarmigan (*Lagopus rupestris*), highly esteemed for food, is stated by Rink to be "pretty common, being in summer almost everywhere met with at heights of from one thousand to two thousand feet, while in winter flocks of them are sometimes seen close to the houses. Very few are caught by snares, almost the whole of them being shot with fowling pieces, and mostly by persons who are engaged and provided by the

* Flower and Lydekker, Introduction to the Study of Mammals, 1891, pp. 267, 268.

Europeans with the necessary implements for taking them. The whole annual production of ptarmigan may be rated at about 12,000 on an average." Concerning the raven (*Corvus corax*), nearly cosmopolitan in range, but rare or absent throughout the eastern half of the United States, Rink says : "The *ravens* are assiduous guests in every settlement when the country is covered with deep snow, and when even berries, which otherwise seem to constitute part of their food, are difficult to be got. They then become almost tame, and follow people carrying seal flesh or blubber in order to snatch the snow that may have imbibed some dropping blood or oil. Their flesh is eaten by few persons, and it is generally not considered worth while to shoot them."

Of the sea fowl, which are especially useful to the Eskimos for food and clothing, Dr. Rink writes :

In summer, swarms of sea fowl are scattered over the whole extent of the coast. It is well known that for the purpose of escaping their enemies some of them, especially the eider ducks, breed in small and low outer islands, while others inhabit precipitous rocks, the so-called bird cliffs. The breeding places of the *eider ducks* are limited to certain clusters of islets, which are regularly visited by the natives in June and July in search of eggs and down. The eggs, even when containing little chickens, are not at all offensive to their heroic taste. The same recklessness with which the natives now waste these eggs has no doubt been shown by them in ancient times ; for this reason it is rather surprising that a more constant decrease in the production of down does not appear to have set in before the last twenty years, the average quantity exported having diminished from fifty-six hundred to two thousand pounds [of raw eider down] yearly during this period. The only probable reason may be found in a more general persecution of the bird itself having of late been added to the devastations of the nests.

Compared with eider ducks, the *sea fowl which inhabit the cliffs* are much less subject to have their nests and eggs taken.

These precipitous walls rising abruptly from the sea to a height
of one thousand or two thousand feet, or even more, with all their
protruding edges and their holes and fissures crowded with birds,
offer a curious sight on account of the immense number of their
eathered inhabitants, and the enormous size of the beetling rocks
when regarded from a boat a few hundred feet distant from the
shore. The appearance of such rocks is generally illusive to the
eye by appearing nearer and lower than they really are, for which
reason the size of the birds at the same time will appear too small.
On seeing the innumerable white patches with which the gloomy
walls are dotted over, we are reminded of snow, while some single
birds which happen to be soaring in the air present the appearance
of down or feathers borne by the wind. But on firing a gun, or on
some other sign being given, the white spots begin to move; in a
moment thousands of birds swarm over one's head, filling the air
with their discordant cries, and the beholder then first receives a
correct impression of the true size and height of the wall inhabited by them. Some of the bird cliffs, and especially those of the
farthest north, contain different species of sea fowls ranged over
one another, the auks occupying the lowest part, the kittiwakes
being the chief inhabitants of the centre, and the gulls inhabiting
the most inaccessible heights. The Greenlanders know nothing
about those peculiar contrivances made use of in other countries
to get at the nests: they merely step from their skiff upon the
rock and climb wherever they are able to find the least footing . . .

In winter all the sea fowls migrate to the south, where open
water may be found, and there, south of 66 north latitude, they
afford profitable hunting to the natives during a season when sometimes they have nothing else but fish for food; the feathers furnish
them with an article of trade, and the skins, with the feathers or
down still adhering to them, form excellent clothing, being at the
same time light and warm. Some of them, distinguished for their
colour and softness, even yield a valuable fur for the European
market in the shape of coverlets or of articles for ladies' dress.
By far the greater part of these birds consist of auks and eider
ducks, and although a great many of them are now shot, they are
still chiefly taken from the kayaks by means of the bird spear.
Sea-fowl jackets are, on account of their lightness, much used by
the kayakers. The whole amount of sea fowls annually killed

may be rated at twenty thousand eider ducks and other larger kinds and more than fifty thousand auks and other smaller kinds. The eggs yearly taken, chiefly those of eider ducks, may be estimated at more than three hundred thousand.

The great auk (*Alca impennis*), formerly inhabiting the northeastern coasts of our continent from Massachusetts northward, also Greenland, Iceland, and the northwestern shores of Europe, has probably now become everywhere extinct, being last known in Iceland in 1844, and a solitary specimen in Labrador in 1870. Unable to fly, these large birds fell an easy prey to man's hunger, and their gradual extermination was well advanced on our northeastern coast before the coming of European colonists, as is shown by the plentiful bones of the great auk in the aboriginal shell mounds.

Lütken's catalogue of the fishes of Greenland includes 79 species. Notes by Dr. Rink, on the species most valuable for food, are as follows:

Sharks (Somniosus microcephalus) are found roaming about everywhere, and will soon appear wherever a large carcass is found or a plentiful capture of seals happens to take place. Those that are caught vary in length from six to sixteen feet, and the liver, forming as yet almost the only part retained for use, weighs between twenty and sixty pounds, in rare instances even several times more. This monstrous fish appears almost as indolent and torpid as it is voracious. Curious instances are related of the greediness and regardlessness of danger exhibited by them when crowding round the carcass of a whale, from which they are not to be scared away even by being severely wounded and mutilated. . . . Several modes of fishing them from open boats and by different sorts of hooks and lines have been attempted, but none of them have proved more effectual than the fishery through holes in the ice. This has been done not only with lines or chains, but also by drawing them to the hole merely by means of torchlight, and then taking them with sharp hand hooks, two men being required to haul each of the larger fish up on the ice.

17

The catch being first successfully commenced in a certain spot, sharks will soon be attracted, and it may be continued in the same place for a great part of the winter. The huge carcasses spreading over the ice then accumulate to several hundreds; at some stations, in favourable seasons, even thousands have covered the ice, attracting ravens, foxes, and especially dogs. But to the latter this frozen shark's flesh has proved obviously unwholesome when swallowed in large quantities and forming their only food for any length of time. It renders them sluggish and torpid and subject to fits of giddiness; on having pulled the sledge a short distance, their ears begin to droop, they tumble from one side to the other, and finally fall into convulsions, and can not be compelled to stir from the spot. The contagious disease [called *piblockto*, by which Hayes and Peary lost the greater part of large teams of sledge dogs] . . . bears a great similarity to this complaint, and as it commenced a few years after the shark fishery had gained its highest pitch, there may be some reason for believing that this disease might have originated from the same source. The bones, being merely cartilage, are considered good eating by the natives, especially after having been kept for a certain time; a little of the flesh is cut into slices and dried, but by far the greater part of the carcasses is thrown away. The flesh has, however, proved to be very rich in oil, and there seems no doubt that its unwholesome qualities merely appertain to it in a raw and particularly in a frozen state. A shark of middle size, weighing about three hundred pounds, contains about one hundred pounds of pure flesh. The number annually captured varies from ten thousand to twenty thousand.

The codfish of Davis Strait [and also of the Canadian and New England fishing banks, as well as of Europe] (*Gadus morrhua*) does not spawn on the shores of Greenland. Spawners are only very rarely caught, and during the winter the cod is wholly absent. Sometimes in spring a great many quite young ones arrive at the inlets between 60 and 61° north latitude, which would seem to suggest that their breeding places were not far off; but they generally make their appearance after June 20th on the fishing grounds, which are situated between 64° and 68° north latitude, at a distance of sixteen miles from the shore, and in July and August resort to the inlets up to about 70° north latitude. With regard to number, the occurrence of codfish on the Green-

land shores is peculiarly variable. Some years, or certain periods
of few years, may prove extremely favourable as regards the
catch, whereas others turn out a total failure. The number annually caught by the natives may be estimated at somewhat about
two hundred thousand fish on an average.

Salmon trout (*Salmo carpio*) occur in the lakes and brooks
and at their outlets along the whole coast, but their capture will
hardly ever attain any importance, because it necessitates people
who undertake it to stay in remote places during the best part of
the summer time. A few are caught in nets to be exported, while
the greater number are either harpooned or speared from the
river sides or from weirs built across the rivers.

The *nanartak*, or *larger halibut* (*Hippoglossus vulgaris*), occurs on the banks as well as in different places outside the islands
up to 70° north latitude in depths of from thirty to fifty fathoms.
Of late the capture of this fish has become an object of commercial speculation, and foreign ships, chiefly American, have been
engaged in it, apparently with better success than that of the cod
fishery. A halibut of this species weighs from twenty to a hundred pounds, and its flesh is fat and much valued. Superior in
taste as well as fatness is the smaller halibut, or *kaleralik* (*H.
pinguis*), which is angled for in the ice fiords at depths of about
two hundred fathoms. The "red fish" (*Sebastes norvegicus*),
found only in certain though pretty numerous grounds south of
68° north latitude, is hauled up from depths of one hundred and
twenty to one hundred and eighty fathoms, and its flesh is likewise rich in oil, which occasionally, in times of want, is extracted
by boiling, and used instead of blubber. The *nepisak* (*Cyclopterus
lumpus*), perhaps the fattest of the Greenland fishes, goes inshore
in April and May for the purpose of spawning, and forms at this
season, during a couple of weeks, the chief food in certain places,
the spawn being also collected and considered a dainty.

The *angmagsat*, or *capelins* (*Mallotus villosus*), has from times
of old yielded the most profitable fishery to the Greenlanders, and
may, in a dried state in winter time, frequently be said to have
constituted the daily bread of the natives. They are shovelled on
shore by means of small nets by women and children, and spread
over the rocks to dry during four weeks of May and June, when
they crowd to the shores of inlets south of 70° north latitude to
spawn. This fishery has now considerably decreased, but may

still be considered to yield one million and a half pounds weight or more of undried fish yearly.

Lastly, we have to mention certain kinds of fish which, although inferior in quality, are nevertheless of inestimable value to the improvident population on account of their being so widely spread, and generally to be had at a season when other provisions are most scanty. These are the *ovak*, or smaller cod (*Gadus ovak*), the *frogfish*, or *kanajok* (*Cottus scorpius*), and the *misarkornak*, or smallest cod (*Gadus agilis*). The two former are found together almost everywhere, though gradually decreasing in number toward the north, while, on the other hand, the latter seems to begin about the middle part of the coast, increasing so as to become abundant in the furthest north. If to the capture of these fishes we finally add the gathering of common *mussels*, which are generally to be found at low water where the shore is not totally closed up with ice, besides . . . seaweeds, we have enumerated the several means by which the final shortcomings of the yearly housekeeping of a Greenland family are made up for, and which almost every year, in some place or other, become the means of saving the people from direst want, and not unfrequently from death by starvation.*

Mörch catalogues 226 species of molluscs belonging to western Greenland, of which he had seen 176 species, while 50 were reported from records by others. There are 7 species of land univalve shells and 4 of fresh-water shells. Some of the others live on the seashore and along the fiords, between the levels of the high and low water of tides, which there have a range of about ten feet; but the great majority are obtained by dredging in depths ranging from the low-water line down to 1,750 fathoms, or two miles. Many of the marine shells, however, are also obtained from the stomachs of the cod and other fishes which feed on them.

---

* Danish Greenland: Its People and its Products, 1877, pp. 131–135.

The insects of Greenland in Schjödte's list, made in 1857, number 124 species, including 21 beetles, 29 species of butterflies and moths, and 48 species of flies and mosquitoes. Mosquitoes are encountered very plentifully by nearly all who visit Greenland. During the warmth of the arctic summer of constant sunlight they become sometimes far more abundant than even in the northern United States and Canada, where, both in the woods and on the prairies, they often astonish and harass those who have lived chiefly in towns or in the older and longer cultivated parts of the country. Nansen writes of his experience one memorable morning with an exceptional abundance of mosquitoes on the east coast of Greenland, previous to his setting out on his journey across the ice sheet.

I woke to find myself scratching my face vigorously, and to see the whole tent full of mosquitoes. We had begun by taking great pleasure in the company of these creatures on the occasion of our first landing on the Greenland coast, but this day cured us completely of any predilections in that way; and if there is a morning of my life on which I look back with unmitigated horror, it is the morning which I now record. I have not ceased to wonder, indeed, that we retained our reason. As soon as I woke I put on my clothes with all speed and rushed out into the open air to escape my tormentors. But this was but transferring myself from the frying pan to the fire. Whole clouds of these bloodthirsty demons swooped upon my face and hands, the latter being at once covered with what might well have passed for rough woollen gloves.

But breakfast was our greatest trial, for when one can not get a scrap of food into one's mouth except it be wrapped in a mantle of mosquitoes, things are come to a pretty pass indeed. We fled to the highest point of rock which was at hand, where a bitter wind was blowing, and where we hoped to be allowed to eat our breakfast in peace and enjoy the only pleasure of the life we led. We ran from one rock to another, hung our handkerchiefs before our faces, pulled down our caps over our necks and ears, struck out and beat the air like lunatics, and, in short, fought a most desperate encounter against these overwhelming odds, but all in vain. Wherever we stood, wherever we walked or ran, we carried with us, as the sun his planets, each our own little world of satellites, until at last in our despair we gave ourselves over to the tormentors, and, falling prostrate where we stood, suffered our martyrdom unresistingly while we devoured food and mosquitoes with all possible despatch. Then we launched our boats and fled out to sea. Even here our pursuers followed us, but by whirling round us in mad frenzy tarpaulins and coats and all that came to hand, and eventually by getting the wind in our favour, we at last succeeded in beating off, or at least escaping from, our enemy.*

In the remaining and lower classes of the Greenland fauna some idea of the extensive observations and researches of naturalists visiting the country or examining collections from it may be obtained by the following totals of the species enumerated in Dr. Rink's work : From Lütken's tabulation, in the class of the Tunicata, 13 species of the simple ascidians are known, while the compound ascidians, though rather numerous, have not been studied and identified ; of Polyzoa, also according to Lütken, 64 species are known ; of arachnids (spiders and their allies), according to Schjödte, 13 species ; of crustacea, according to Reinhardt and Lütken, 202 species ; of the Annulata (annelid worms), according to Lütken, who also is authority for all the following, 133 species ; of Entozoa (intestinal worms), 62 species ; Echinodermata (starfishes, etc.), 34 species ; Anthozoa (sea anemones, etc.), 15 species ; acalephs (jellyfishes, etc.), 33 species ; and sponges, 28 species.

---

* The First Crossing of Greenland, 1890, vol. i, pp. 396–398.

CHAPTER X.

WHEN Agassiz, in 1840, from his studies of the glaciers of the Alps and of their previous much greater extent, announced his grand generalization that the drift covering northern lands was due to great ice sheets which since have vanished, little was definitely known of the existence of the ice sheets which now have been ascertained to cover the greater part of the antarctic continent and of Greenland. Reports of scanty observations suggesting the existence of these great continental ice-fields in both the arctic and antarctic regions had been brought to Europe, but no demonstrative explorations had yet proved that the snow and ice were of vast extent and thickness. Not only has the glacial theory of Agassiz received continually cumulative support by its affording adequate explanations, derivable from no other source, for many characteristics of the drift and for its various phases of deposition and the diversity of conditions attending its origin, as these have been gradually made known by the progress of geological surveys, but also the theory has found exemplification by now existing ice sheets, that of Greenland being found to have an area approximately one fourth as large as the Pleistocene ice sheet of Europe, while that surrounding the south pole is somewhat more extensive than the old ice sheet

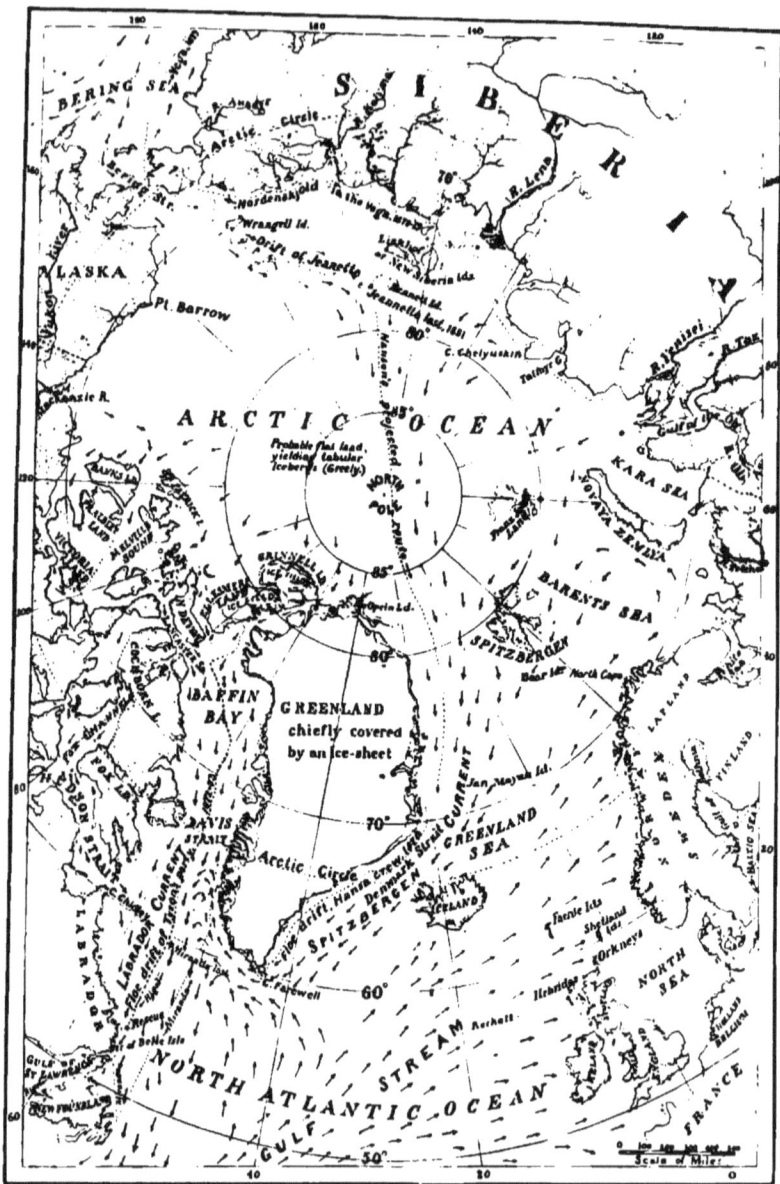

of North America, which covered about 4,000,000 square miles.

It was not long, however, after the glacial studies of Agassiz had set all geologists to new and fruitful thinking and observing for explanations of the drift that the voyages of Ross, in 1841 and 1842, brought tidings of the border of the antarctic ice sheet terminating in the ocean with frontal perpendicular cliffs 150 to 200 feet or more in height, along which, in latitude 77° 45' to 78° south, he sailed 450 miles eastward from Mounts Erebus and Terror, finding only one place low enough to allow the upper surface of the ice to be viewed from the mast-head. There it was a plain of snowy whiteness, reaching into the interior as far as the eye could see.

These and other voyages to the antarctic regions show that land, chiefly covered by a vast *mer de glace*, extends to a distance of 12 to 25 degrees from the south pole, having an area, according to Sir Wyville Thomson, of about 4,500,000 square miles. Whether the antarctic ice sheet covered an equal or greater extent in the Pleistocene period, contemporaneous with the glaciation of now temperate regions, we have no means of knowing. That the ice plain has a considerable slope from its central portions toward its boundary is shown by its abundant outflow into the sea, by which its advancing edge is uplifted and broken into multitudes of bergs, many of them tabular, having broad, nearly flat tops. As described by Moseley in Notes by a Naturalist on the Challenger, these bergs give strange beauty, sublimity, and peril to the Antarctic Ocean, upon which they float away northward until they are melted. Many parts of the borders of the land underlying this ice sheet are low and almost level, as is known by the flat-topped and horizontally stratified bergs, but some other areas are high and mountainous.   Due south of New Zealand the volcanoes Terror and Erebus, between 800 and 900 miles from the pole, rising respectively about 11,000 and 12,000 feet above the sea, suggest that portions or the whole of this circumpolar continent may have been recently raised from the ocean to form a land surface, which on account of its geographic position has become ice-clad.

Inside its border of mountains, Greenland is enveloped by an ice sheet which has a length of about 1,500 miles, from latitude 60° 30′ to latitude 82°, with an average width of almost 400 miles, giving it an area of about 575,000 square miles.*   On the east this ice sheet in some places stretches across the mountains, and the coast consists of its ice cliffs ; and on the west glaciers flow from the inland ice through gaps of the mountains to the heads of the many fiords and bays, where the outflowing ice is broken into bergs of every irregular

* Dr. Henry Rink, in 1877, estimated the area of Greenland as 512,000 square miles, of which he considered the inland ice sheet to cover 320,000 square miles.   Dr. John Murray, in the Scottish Geographical Magazine for January, 1888 (vol. iv, p. 7), computes the area of Greenland, according to maps, to be 914,550 square miles ; and he estimates its mass above the sea level to be 556,350 cubic miles, the average elevation being regarded as about 3,200 feet.   Lieut. R. E. Peary, in the Bulletin of the American Geographical Society (vol. xxiii, p. 159, June 30, 1891), estimated the area of Greenland as 740,000 or 750,000 square miles : and he stated that " no less than four fifths of this area, or 600,000 square miles, . . . is covered by the inland ice."   Later, in Johnson's Universal Cyclopædia (vol. iv, 1894, pp. 25–27), Peary gives the whole area as about 500,000 square miles, of which the ice sheet is thought to occupy " over 400,000 square miles."   Measurements on our map of Greenland show the whole area of the country to be very nearly 680,000 square miles, and of the ice sheet about 575,000 square miles.

shape and borne away by the sea. One of these ice streams, discovered and named by Kane the Humboldt Glacier, is 60 miles wide where it enters Peabody Bay, above which it rises in cliffs 300 feet high. The general boundary of the ice sheet upon the mountains and plateaus of the border of Greenland, excepting the out-flowing valley glaciers, has usually a height of 1,500 to 2,000 feet above the sea; and thence the ice surface gradually rises to the great altitude of 8,000 to 9,000 feet, or more, in its central part. Near its boundary the ice sheet, there undergoing more ablation or superficial melting than snow accumulation, is commonly much intersected by crevasses, due to its flow over a surface of varying gradients, and is made uneven by the small pin-nacles and ridges of its irregular melting, so that it is very difficult of ascent for the first few miles. Farther within the ice area, it has been found by Hayes, Nor-denskjöld, Nansen, and Peary to have a very even snow-covered surface, well adapted for travel with sledges and snowshoes. This great central region of the inland ice, with its snow covering, is the analogue of the *névé* fields or gathering grounds of the ice of all Alpine glaciers. Instead of an ice surface, it is wholly a vast snow *névé* field after the comparatively narrow peripheral zone of ablation, ice ridges, and yawning crevasses is passed.

In a valuable chapter of Nansen's First Crossing of Greenland, which is also largely reprinted in the Bulle-tin of the American Geographical Society (vol. xxiii, pp. 171–193, June 30, 1891), the various journeys on the Greenland ice sheet, and observations of its margin, are brought under review; and from this and other sources the following notes, arranged in chronologic order, are derived :

An ancient Norse treatise, called the *Kongespeilet*,

or " King's Mirror," written apparently in the thirteenth
century, portrays, as quoted by Nansen, the condition of
the interior of Greenland, so far as the early Icelandic
immigrants had explored the country:

Seeing that thou hast asked whether the land is free of ice or
not, or whether it is covered with ice like the sea, thou must know
that that part of the country which is bare of ice is small, and that
all the rest is covered with ice, and that people therefore know not
whether the country be large or small, seeing that all mountains
and all valleys are covered with ice so that one can nowhere find
an opening therein. And yet it would seem most credible that
there should be an opening either in the valleys that lie among
the mountains or along the shores, by which animals can find
their way, since otherwise animals could not wander hither from
other lands unless there be an opening in the ice and land free
from ice. But oftentimes have men tried to come up into the
land upon the highest mountains that be found there and in divers
places, in order to see round about them, and to discover if perad-
venture they could find land which was bare of ice and habitable,
and they have nowhere found such, but only that on which people
now dwell, and for a little way along the very shore.

After the recolonization of Greenland in 1721 by the
Norwegian missionary Hans Egede and his companions,
only two years passed before the enterprise of the Bergen
Company, under whose patronage for the exploration of
the land and development of commercial relations with
the Eskimos the new mission and colony were established,
found expression in the following letter of instructions
sent to Egede:

It seems to us quite advisable, if indeed the thing has not been
done already, that a party of eight men should be told off to
march through the country, which, according to the map, would
appear to be only from eighty to a hundred miles across at its
narrowest part, for the purpose of reaching, if it be possible, the
east side, where the old colonies have been, and on their way to
look out for forests and other things. If this is done, as we

should much like, the thing must be undertaken in the early summer; and, furthermore, the men must be provided each with pack, provisions, and gun, as well as with a compass, in order that they may be able to find their way back again; and, thirdly, the men of the party must both look out warily for the attacks of savages, in case they should fall in with any on the way, and must also make all possible observations, and wherever they pass must raise piles of stones upon high places, which will serve as marks both for this and future occasions.

Nansen well remarks of this recommendation that it is "an amusing instance of the achievements of colonial policy under the guidance of geographers of the study and easy chair." Egede replied that the maps were unreliable, and that to traverse the country to the east coast would be difficult or impossible, on account of the high cliffs and the mountains of ice and snow. In 1727 a letter sent to Europe from Godthaab stated that "following the backbone or central ridge of the country from south to north was an appalling tract of ice, or mountain covered with ice."

Undaunted by this report, the home Government of Denmark, which had succeeded the commercial company of Bergen, Norway, in the support of the Greenland colony, in 1728 instructed Paarss, who had been appointed Governor of Greenland, "to spare no labour or pains, and to allow himself to be deterred by no danger or difficulty, but to endeavour by all possible means, and by one way or another, to cross the country . . . for the purpose of learning whether there still exist descendants of the old Norwegians; what language they speak; whether they are Christians or heathens, as well as what method of government and manner of life prevail among them; . . . and what is the true nature of the country; whether there is forest, pasturage, coal, minerals, or other things of the kind; whether there are horses, cattle, or other animals suited to the service of man."

With the expectation that the expedition could ride on horseback across Greenland, eleven horses were sent from Denmark, but five died on the voyage, and the others "soon perished from hunger and hardships in Greenland." After a winter of the utmost distress and discontent in Paarss's colony, at the newly founded station of Godthaab, he set out, with seven others, April 25, 1729, on the proposed expedition, sailing to the head of Ameralik Fiord (which in 1888 was the end of Nansen's journey across Greenland), and thence marching two days inland to the edge of the ice sheet. The Governor described his efforts there as follows:

> When we had ascended this and advanced upon it for two hours at our great peril all farther progress was denied us by reason of the great chasms which we found thereon. . . . As soon as we saw that no farther advance was possible, we sat ourselves down upon the ice, with our guns fired a Danish salvo of nine shots, and in a glass of spirits drank the health of our gracious king on a spot on which it had never been drunk before, at the same time paying to the "ice mountain" an honour to which it had never before attained: and after we had sat and rested ourselves for about one hour we turned back again.

The large boulders which were seen on the border of the ice were thought to have been swept upon it "by great and violent winds and tempests, which there have incredible fury."

Nansen observes that this expedition, though rather ridiculous in its achievements, "can not have failed to have considerable effect at home in Copenhagen, seeing that the next expedition organized by the Danish Government was not sent out till 1878, or a hundred and fifty years later."

A more extended exploration of the ice border was
accomplished by Lars Dalager, a merchant of Frederikshaab, in September, 1751, reaching large nunataks a
few miles inside the ice sheet at a distance of about
twenty miles inland from the termination of the great
glacier known as "Frederikshaabs Isblink." Of his
journey Dalager wrote as follows:

My errand was only to divert myself with my gun, but on this
occasion it was not long before I had resolved to set out on a
journey across the "ice mountain," to "Osterbygd," to which
determination I was led by a new discovery made . . . by a
Greenlander, who had been so high up while out hunting that he
could see distinctly, as he said, the . . . mountains on the eastern
side. This moved me, as I have said, with a desire at least to see
the land, like Moses of old, and I took with me the aforesaid man
and his daughter, together with two young Greenlanders. We
set out upon our journey after having already advanced thus far
into a fiord by the southern side of the glacier. . . . In the morning we committed ourselves to the ice, purposing to reach the first
mountain top, which lies in the middle of the ice field, and which
was five miles distant from us. So far the ground was as flat and
smooth as the streets of Copenhagen, and all the difference that I
could see was that here it was rather more slippery, but on the
other hand one had not to wade out to the sides in the slush in
order to avoid being overthrown by the posting horses and car-
riages. . . . [Thence they went forward the next day] to the
uppermost mountain on the ice, called Omertlok, to which it was
also about five miles, but here the ice was very rough and full of
cracks, for which reason it took us seven hours to reach it.

From this nunatak they looked across the ice northeastward some fifteen miles to other peaks, which the
Eskimo hunter and Dalager supposed to be near the
east coast, but which are now known as Jensen's nunataks (noticed more fully on a later page) and are found
to be situated also near to the western border of the ice
sheet, in comparison with its entire width.

Dalager wished to advance farther, but says: " I was constrained for many reasons to set my face towards home, one being very important, that we were now going no better than barefooted; for, though each of us was provided with two pairs of good boots for the journey, yet they were already quite worn out by reason of the sharpness of the stones and ice. And as the handmaid whom we had in our company had, to our great misfortune, lost her needle, we could get none of our things mended. For this cause we were much embarrassed, though we consoled each other with laughter as we contemplated the naked toes peeping out from the boots." The crevasses of the ice surface seemed to Dalager to oppose no insuperable obstacles to a journey across it to the east coast; but on other accounts he thought the crossing impracticable, because "one can not drag as much provision as one should reasonably be provided with on such a march, and, further, on account of the intolerably severe cold, in which I think it all but impossible that any living creature could exist, if he were to encamp for many successive nights upon the icefield. . . . I can say that of all the bitter winter nights on which I have camped on the ground in Greenland, none have so much distressed me by reason of the cold as these nights early in the month of September."

Passing the slight observations of the inland ice by Fabricius and Giesecke, we find a long interval of about fifty years which brought no additional knowledge on this subject. The next important author was Dr. Henry Rink, whose elaborate work on Greenland, resulting from many years of residence and exploration there, was published first in Denmark, in 1857, and twenty years later was revised and published both in Danish and English. This work, following upon the glacial theory of Agassiz and the new zeal with which the drift of Europe and North America was being studied, greatly interested physicists and geologists in its descriptions and map of the Greenland ice sheet.

No exploration of the ice sheet to any considerable distance from its margin had been done when Rink, describing the relationship between the western ice-free land belt, with its deeply indenting fiords, and the great inland ice enveloped area, wrote :

Wherever these fiords have been followed to their terminations, and an attempt has been made to penetrate the regions beyond, or to attain a view from the adjoining heights, the country has been found to exhibit the same continuous waste of ice. When visiting the southernmost portion of the mainland in the environs of Cape Farewell, near the latitude of Christiania, we meet with the very same hindrance as on the coast a thousand miles farther north. On entering these southern fiords we are first struck with the luxuriant vegetation, gradually increasing toward their termination. The charming scenery of the verdant valleys and slopes here displayed leads the traveller to suppose that a few miles still farther inland the country will be covered with wood, and change its arctic character. So far from this, wherever we follow a fiord to its source and try to proceed farther in the same direction by land, we are suddenly arrested by a wall of ice rising abruptly from the ground, which in the immediate neighbourhood produces vegetation. But if we subsequently, in order to find some other passage, ascend a neighbouring hill, thinking that the ice wall probably belongs to some glacier of a limited extent, we see that it forms the unbroken edge of an elevated icy plateau, sloping gently down toward the sea and occupying the whole interior. As far as this plain can be overlooked from the heights of the outer land, or has been travelled over (to a distance of twenty miles from its nearest seaward border), it only attains a height of a little more than two thousand feet, but must be supposed to still rise very gradually toward the wholly unknown interior, where no human foot as yet has trodden. This elevation is much less than that of the outlying headlands which sever the inland ice from the open sea, which to the north and the south frequently attain a height of from three thousand to four thousand feet. This circumstance, combined with its uniformity and other reasons, . . . contradicts the opinion that, like other glaciers or *mers de glace*, it rests upon a high table-land. On the contrary, its probable thickness and extent may rather be compared with an inundation that has overspread the interior in the course of ages, only for some reason or other kept within a certain limit toward the sea. The analogy to an inundation is furthermore in accordance with occasional small insular hills or rocks, called nunataks by the natives. These, however, seldom rise from the uniform horizontal surface, representing the still emerging mountain tops of the vanished land, which, on the whole, at least within the first fifty or one hundred miles from its western border, seems to have been low in comparison with the bold headlands which project as its continuation seaward.*

Incited by Rink's descriptions of the inland ice, and by the work of geologists in exploring the character and extent of the North American and European drift deposits, and in seeking to explain their origin, numerous observers from the years 1859 and 1860 onward have added much to what was previously known of this great ice sheet.

Schaffner, in 1859, and Rae, near the end of October in the following year, made short excursions of reconnoissance on the border of the ice sheet in the vicinity of Julianshaab (latitude 60° 40′), near the south end of Greenland.

Twelve hundred miles farther north, at nearly the same time with Rae's expedition, a more notable journey on the ice east of Port Foulke (latitude 78° 18′) was made by Dr. I. I. Hayes, starting October 22d and occupying six days. On the second day Hayes and his

---

* Danish Greenland. pp. 41, 42.

five companions reached the ice border and scaled its steep and much crevassed frontal slope, and advanced thence an estimated distance of five miles. "As we neared the centre of the glacier," Hayes writes, "the surface became more smooth, and gave evidence of greater security. The great roughness of the sides was no doubt due to an uneven conformation of that portion of the valley upon which the ice rested." For this distance the angle of ascent was estimated as 6°; but the next day, when the party travelled "thirty miles," the gradient was found to decrease to about 2°. Dr. Hayes further writes of this journey, which was nearly due east, along or near to the parallel of 78°, upon the part of the ice sheet immediately north of Inglefield Gulf, being the same tract which Peary crossed in 1892, when setting out on his great journey to Independence Bay:

From a surface of hard ice we had come upon an even plain of compacted snow, through which no true ice could be found after digging down to the depth of three feet. At that depth, however, the snow assumed a more gelid condition, and, although not actually ice, we could not penetrate farther into it with our shovel without great difficulty. The snow was covered with a crust through which the foot broke at every step, thus making the travelling very laborious. [The distances, therefore, and probably also the altitude stated later, seem overestimated.]

About twenty-five miles were made the following day, the track being of the same character as the day before, and at about the same elevation; but the condition of my party warned me against the hazard of continuing the journey. The temperature had fallen to 30° below zero, and a fierce gale of wind meeting us in the face drove us into our tent for shelter, and, after resting there for a few hours, compelled our return. I had, however, accomplished the principal purpose of my journey, and had not in any case intended to proceed more than one day farther at this critical period of the year.

. . . The temperature fell to 34° below zero during the night;

and it is a circumstance worthy of mention that the lowest record of the thermometer at Port Foulke during our absence was 22° higher. . . . The storm steadily increased in force, and, the temperature falling lower and lower, we were all at length forced to quit the tent, and in active exercise strive to prevent ourselves from freezing. To face the wind was not possible, and shelter was nowhere to be found upon the unbroken plain. There was but one direction in which we could move, and that was with our backs to the gale. . . . Our situation at this camp was as sublime as it was dangerous. We had attained an altitude of five thousand feet above the level of the sea, and we were seventy miles from the coast, in the midst of a vast frozen sahara, immeasurable to the human eye. There was neither hill, mountain, nor gorge anywhere in view. We had completely sunk the strip of land which lies between the *mer de glace* and the sea; and no object met the eye but our feeble tent, which bent to the storm. Fitful clouds swept over the face of the full-orbed moon, which, descending toward the horizon, glimmered through the drifting snow that whirled out of the illimitable distance, and scudded over the icy plain—to the eye in undulating lines of downy softness, to the flesh in showers of piercing darts.

Our only safety was in flight: and, like a ship driven before a tempest which she can not withstand and which has threatened her ruin, we turned our backs to the gale: and, hastening down the slope, we ran to save our lives. We travelled upwards of forty miles, and had descended about three thousand feet before we ventured to halt. The wind was much less severe at this point than at the higher level, and the temperature had risen twelve degrees. Although we reposed without risk, yet our canvas shelter was very cold: and, notwithstanding the reduced force of the gale, there was some difficulty in keeping the tent from being blown away.*

On June 19, 1867, Edward Whymper, the distinguished English traveller and mountain climber, ascended the border of the inland ice from the Ilordlek Fiord, near latitude 69° 30', about twenty miles north of

---

* The Open Polar Sea, 1867, pp. 132–135.

Jakobshavn. This was a reconnoissance with the intention of learning the best place for beginning a more extensive sledge journey which he hoped to make later in the same summer. Whymper's first observations and his subsequent fruitless endeavours are summarized from his narrative, by Nansen, as follows:

The first view showed the surface of the "inland ice" to be much smoother and far less formidable than had been expected. The party ascended it and advanced without difficulty, finding the snow harder and better to walk upon the farther they went. When they had pushed in some six miles, and reached a height of about fourteen hundred feet, and the surface appeared to them to be equally good as far inward as they could see, they considered that the object of the excursion had been attained, and that there was nothing to be gained by advancing farther. They were convinced that the snowfield was eminently fitted for dog sledging, and the Eskimos declared that they could easily drive thirty-five or forty miles a day. They all turned back with the best hopes of success, "for there appeared to be nothing to prevent a walk right across Greenland."

However, as at Iiordlek the ice does not quite come down to the water's edge. Whymper determined to look for a suitable spot where this was the case, so that he might take to the ice at once and avoid the transfer of his baggage over land. So between June 24 and 27 he made another excursion to the edge of the icefield, this time to "Jakobshavns Isfiord," as it is called, which lies to the south of the colony. Here, however, the ice was so fissured and rough that any transport by means of dog sledges would have been impossible, and therefore the spot which they had first visited was decided upon as the starting point of the expedition.

A number of preparations were, however, necessary, and in his attempt to carry out these Whymper was met by difficulties which proved almost insuperable. . . . Most of these obstacles were overcome in one way or another, and on July 20 the expedition was ready to start. The party consisted of five members, Eskimos and Europeans, in addition to Whymper himself, one of the latter being the English traveller, Dr. Robert Brown. Two days or so were spent in carrying the baggage up from the fiord to the edge of the ice, and three more in waiting for more favourable weather.

Meanwhile Whymper ascended one of the neighbouring heights to obtain a view over the ice, and was most unpleasantly surprised to find that the surface had completely changed its aspect. When he had seen it a month before there had been a covering "of the purest, most spotless snow," but this had now melted away "and had left exposed a veritable ocean of ice, broken up by millions of crevasses of every conceivable form and dimensions." . . . However, on July 26, as the weather was now better, an attempt was made to push over the ice eastward. But after advancing for a few hours, and having covered only a couple of miles of ground, the party were brought to a standstill by the breaking of one of the runners of a large sledge, by the splitting of another on one of the smaller, and the general dilapidation of the rest, owing to the rough treatment to which they had been exposed.

. . . The result of the visit to Greenland was that Whymper's belief in the existence of bare land in the interior of the continent was considerably shaken, and in 1871, in his book Scrambles among the Alps, he writes, "The interior of Greenland appears to be absolutely covered by glacier between 68˚ 30' and 70˚ north latitude." . . . He also estimated the height of the most distant part of the "inland ice" within view at "not less than eight thousand feet," an elevation which, though somewhat too high, can not be very far from the truth.*

Mr. Whymper again visited Greenland in 1872, and from the summit of a mountain 6,800 feet high, near the large Umanak Fiord (latitude 71°), he saw eastward "a straight, unbroken crest of snow-covered ice, concealing the land so absolutely that not a single crag appeared above its surface." Sighting with a theodolite to this horizon, he estimated its height to be "considerably in excess of 10,000 feet." He also concluded "that

* Nansen's First Crossing of Greenland, vol. i, pp. 474-479.

the whole of the interior, from north to south and east to west, is entirely enveloped in snow and ice."

The first journey of Baron A. E. Nordenskjöld on the Greenland ice sheet, accompanied by Dr. Berggren and two Eskimos, was July 19th to the 25th, 1870, starting from the head of Aulatsivik Fiord, near latitude 68° 20', and advancing nearly due east an estimated distance of about 35 miles upon the inland ice,* to the altitude of 2,200 feet. Large streams on the ice surface were encountered, "which could not be crossed without a bridge." After flowing some distance, however, these streams, produced by the superficial melting, usually plunged into deep "glacier wells" or *moulins.* Small lakes and pools of water were also found on the ice, into which rivulets flowed, while their waters percolated downward from the lake bottoms. "When one laid the ear down on the ice, one heard from all sides a peculiar subterranean murmur from the streams inclosed below, while now and again a single loud cannonlike report announced the formation of some new crevasse." Beyond the point of turning back, "the inland ice continued constantly to rise toward the interior, so that the horizon toward the east, north, and south was terminated by an ice border almost as smooth as that of the ocean." The temperature at night was a little below freezing, but in the middle of the day it rose to 45° or 46° F., near the *névé* surface, and even to 75° or 85° in the sunshine. "During the whole of our journey on the ice," writes Nordenskjöld,† "we constantly enjoyed fine

---

* This distance was estimated by Nordenskjöld to be "about 80 miles," but Nansen's estimate, as here given, is more probable.

† Geological Magazine, I, vol. ix, pp. 303–306, 355–368, July and August, 1872.

weather, frequently there was not a single cloud visible in the whole sky."

Concerning long past and recent fluctuations in the extent of this part of the ice sheet, Nordenskjöld remarks : " The inland ice, in former times, evidently covered the whole of Aulatsivik's Fiord, together with the surrounding valleys, mountains, and hills. The ice has accordingly, during the last thousand or hundred thousand years, considerably retired. Now, on the contrary, its limit in these parts is advancing, and that by no means slowly."

A fine, gray powder, called " kryoconite," which was believed by Nordenskjöld to be cosmic dust, was found on the ice ; but analyses indicate that this is dust blown from the mountains of the coast, and it does not occur in noticeable amount, according to Nansen, on the eastern portion of the ice sheet, where his ascent was made upon ice bordered by only little bare land. On account of the sun's warmth, the kryoconite on certain tracts had sunk into the ice, leading to the formation of cylindrical and bowllike holes from one to two feet deep, which sometimes were so plentiful that it was difficult to find place for one's feet and for walking among them.

During the year 1875 the first exact measurements of the rate of flow of the Greenland glaciers were made by Helland, who visited five ice fiords, besides several small glaciers not extending into the sea, upon the part of the west coast between Egedesminde (latitude 68° 42′) and the fiord of Kangerdlugssuak (near latitude 71° 15′). He also ascended the front of the inland ice near the same place where Whymper had done so eight years before. Helland ascertained that the central part of the front of the great glacier in the Jakobshavn ice fiord (latitude 69° 15′) moved in July at the rate of 64 feet daily. The breadth of this glacier is 14,000 feet, or 2⅔ miles, and its depth in the centre exceeds 1,000 feet. Five years later, the movement of this glacier in a part about 4,000 feet nearer to its side was found by Hammer to vary from 33 to 51 feet daily in March and April. The declivity of its surface is ascertained to be only a half of one degree, or about 46 feet per mile.

The Torsukatak Glacier (near latitude 69° 50′), having a width of about five miles, was found by Helland to move, at its centre, 30 to 32 feet daily ; and a year or two later Steenstrup measured its rate at some distance from the centre, and found it to be from 16 to 25 feet in twenty-four hours.

The Karajat Glacier (latitude 70° 30′), at the head of the Umanak Fiord, with a width of 19,000 to 22,000 feet, or about four miles, moves in summer, according to Steenstrup, 22 to 38 feet a day ; and the Itivdliarsuk Glacier (latitude 70° 45′), 17,500 feet wide, was found to move 46 feet a day in April, and 21 to 28 feet daily in May.

Fastest in rate of advance among all the glaciers of Greenland thus far measured, is the great glacier outflowing into the Bay of Augpadlartok (latitude 73°), near Upernivik, which, according to Ryder's observations in August, 1886, was found to have a velocity of 100 feet in twenty-four hours ; but a measurement at nearly the same point in April showed a progress of only 34 feet daily.

In the district of Julianshaab, near the southern extremity of Greenland, the measurements of three glaciers in 1876, by Steenstrup, Kornerup, and Holm, gave maximum rates of only about 12 feet in twenty-four hours ; but even this is three or four times the maximum rates of the glaciers of the Alps.

These observers, excepting Helland, their pioneer, were sent out for these and other explorations by the Geological and Geographical Survey of Greenland, and their work has been reported in the Meddelelser om Grönland, published in Copenhagen from 1879 onward. Summing up their work on the movement of the glaciers, Rink in 1888 stated that the rate of motion of twenty-five glaciers, or more, terminating in deep fiords, had been accurately determined, the average rate in summer in their central portions being 51 feet a day. During the colder portions of the year they move more slowly, and at all times the rate of motion diminishes from the centre to the sides. "The true home of icebergs," according to Rink, is "the coast between $68\frac{1}{2}°$ and 75° north latitude, which contains all the large ice fiords on the western side that are thoroughly known."*

Helland, in his views looking eastward from five high peaks of the western land tract, between latitudes 69° 10' and 71° 15', "saw only ice, like a great sea, lying at a much lower level than these peaks, but rising slowly inland and forming an undulating sky line. The surface of the glaciers in the fiords was mostly free from stones, except at the margin. At the Jakobshavn Fiord the discharge of ice in July was equal to one large iceberg a day of about 16,000,000 cubic metres. The discharge continues in the winter, as the icebergs set free prove, but at a slower rate. He observes also that the amount of glacier discharged as ice is far less than that which passes out as water beneath the glacier. The mean

---

* H. Rink, The Inland Ice of Greenland, Scottish Geographical Magazine, vol. v. pp. 18-28. January, 1889 (from the Zeitschrift der Gesellschaft für Erdkunde zu Berlin, vol. xxiii, No. 5).

amount of mud discharged by the waters flowing from
six glaciers he found to be, in July and August, 1875,
727 grammes in one cubic metre of water (or in 100,000
grammes very nearly)." *

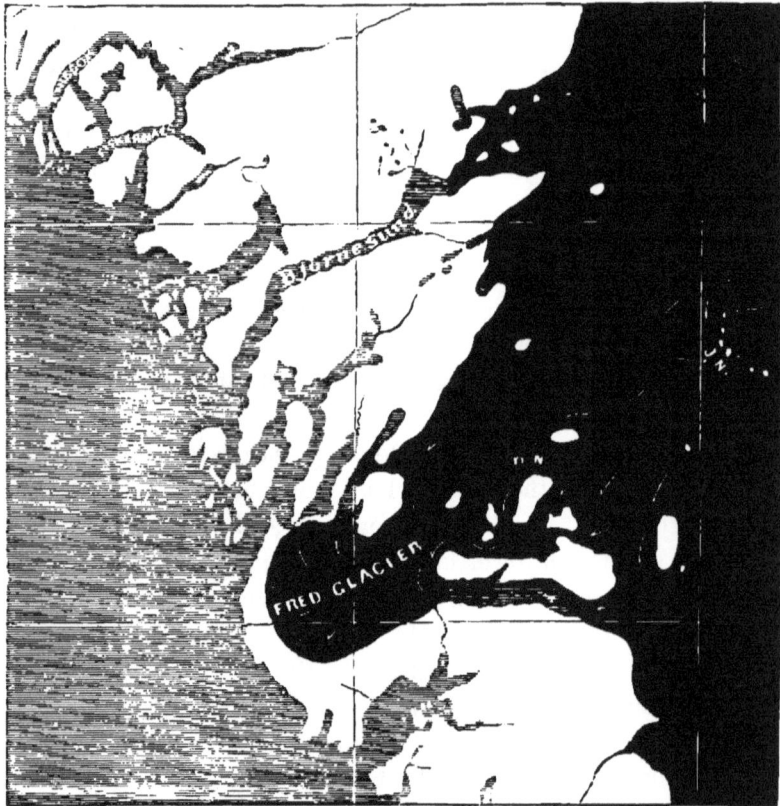

Fig 49.—Explorations of the Greenland ice sheet, in the vicinity of the
Frederikshaab Glacier, by Jensen and Kornerup, 1878. The black part,
ice ; white, land ; shaded, water ; J. N., Jensen's nunataks ; D. N.,
Dalager's nunataks ; white lines on the black, crevasses ; arrows, glacier flow. The two parallels of latitude shown are at 62° 30′ and 63° ;
and the two meridians designated, 49° and 50° W. longitude. (From
Dana, after Jensen.)

* J. D. Dana, in American Journal of Science, III, vol. xxiii,
p. 365, May, 1882, from Helland's reports.

In 1878, Lieutenant J. A. D. Jensen, with Mr. A. Kornerup as geologist, in the service of the Greenland survey, made an important expedition to the nunataks, which Dalager had seen more than a century before, as narrated on a foregoing page. Jensen's party started July 14th from a locality called Itivdlek (near latitude 62° 40'), on the north side of the Frederikshaab "ice

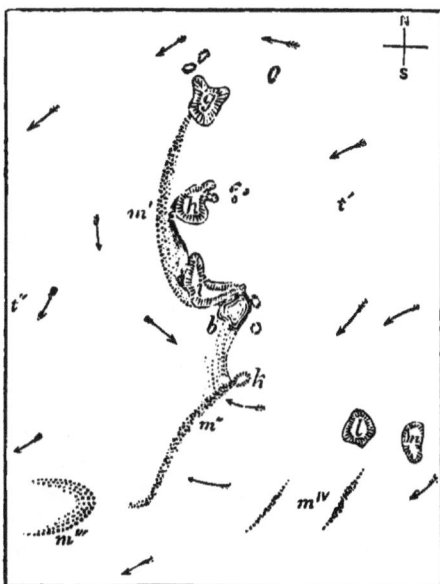

Fig. 50.—Enlarged map of Jensen's nunataks, with the currents and moraines of the surrounding ice sheet. (From Dana, after Jensen.)

blink," and they travelled east-northeasterly about 47 miles, the group of nunataks which they reached, since known as Jensen's nunataks, being near latitude 62° 50' and longitude 49° (Figs. 49 and 50).   The ice was found to be much crevassed, so that travelling was difficult and slow, and, moreover, the party suffered much from snow blindness. The nunataks, which Dalager had supposed to be mountains adjoining the east coast, but which are only about twenty miles from the nearest part of the ice border on the west, were reached on the 24th of July, and the party remained there a week, being detained by a snowstorm.

The observations by Jensen and Kornerup on the relations of the currents of the ice sheet, and on the form of its surface, as influenced by the obstructions of the nunataks or projecting mountain tops, are stated by Dana, from the Danish reports of the Greenland survey, as follows :

The heights above the sea of the four largest were severally, commencing to the north, 5,623 (*g*), 5,184 (*i*), 5,654 (*k*), and 5,580 (*m*) feet. From these peaks, which stand like islands in the sea of ice, moraines of stones and earth (some of the stones 20 feet in their dimensions) extend for 1 to 2½ miles (*m'*, *m''*, *m'''*, *m''''*, Fig. 05); and dust, by the aid of the storm winds, is drifted off for wide distribution over the glacier. The moraines, after a short outside existence, disappeared beneath the ice, the stones dropping down the crevasses that were from time to time opening (the account says) as the glacier moved on. The varying direction of the moraines, and the eddies in the flowing ice, due to the obstructing ridge (of which the nunataks are the peaks), which these directions indicate, are remarkably instructive. The arrows show the inferred direction of movement. The moraine *m'*, 2½ miles long. is made mostly of polished stones, which appear therefore to have travelled far, and not to be the *débris* of the nunataks. The *m''* and *m'''* have no connection with any visible nunatak. At *b* is a lake 824 feet in diameter. and toward it the ice around slopes from a height of 4.900 feet to that of 4.120 feet on its borders. The glacier had a height at *t*, east of the nunataks, of 5,150 feet. The slope of the ice surface for the distance traversed averaged 0° 49', or about 75 feet per mile (1 : 70), and it was evident that the movement of the glacier depended on this slope. Crevasses were numerous along the route transverse to the line of movement as well as longitudinal and radial (see fine lines along the route on Fig. 49); and fresh-water streams were common, and waterfalls also. The largest of Jensen's nunataks consisted mostly of hornblende schist in bold flexures, with mica schist and gneiss.*

In 1880, Dr. N. O. Holst, the Swedish geologist, examined portions of the margin of the Greenland ice sheet, his purpose being chiefly to study its varieties of drift, their modes of occurrence, and the processes of their deposition. He found extensive deposits of both englacial and subglacial drift, respectively characterized by angular and by glaciated stones and boulders. The largest accumulation of superglacial drift, which had been englacial, was observed on the southern edge of the lobe of the ice known as the Frederikshaab ice blink (latitude 62° 30'). The drift covering the ice surface here, as exposed by the ablation or superficial melting, was ascertained to extend along a distance of nearly twelve miles, and to reach half a mile to a mile and a half upon the ice. According to Holst's Swedish report of his observations, summarized in translation by Dr. Josua Lindahl,† the quantity and upper limit of the superglacial drift at this locality are as follows:

Its thickness is always greatest near land, but here it is often quite difficult to estimate its actual thickness, as it sometimes forms a compact covering, only in some fissures showing the underlying ice. This uneven thickness of the moraine cover offers to the ice a proportionally varying protection against the sun. It thus happens that the unequal thawing moulds the underlying surface of the ice into valleys and hills, the latter sometimes rising to a height of 50 feet above the adjacent valley, and being so densely covered with moraine material that this completely hides

---

* American Journal of Science, as before cited, p. 364. The plants collected by Kornerup on these nunataks, as here reported, have been noticed in a preceding chapter (p. 198).

† American Naturalist, vol. xxii, pp. 589–598 and 705–713, July and August, 1888.

the ice core, which, however, often forms the main part of the hill.

Farther in on the ice the moraine gradually thins out. At the locality just referred to the moraine cover, 3,000 feet from land, measured several inches in depth; still the ice was seen in some bare spots. Beyond 4,000 feet from land the moraine formed no continuous cover, and at 8,300 feet it ceased entirely, with a perceptible limit against the clear ice. Only some scattered spots of sand and gravel were met with even a few hundred feet farther in on the ice. Dr. Holst estimated the average thickness of the moraine taken across its entire width near its eastern end at one to two feet. The limit between the moraine cover and the pure ice is always located at a considerable though varying elevation above the edge of the inland ice. In the instance of the above-mentioned moraine it varied between 200 feet and 500 feet.

Terminal moraine ridges, in process of accumulation on the thinned border of the ice, were seen in several places, sometimes, as shown by the following quotation, consisting chiefly of subglacial drift, elsewhere of englacial drift:

The border moraines north of the Arsuk Fiord ice river [latitude 61° 10′] are visible far out on the sea off Ivigtut. Dr. Holst examined one that surrounds the southernmost strip of land at a distance from land of about 2,000 feet. It is not one continuous ridge, but consists of several disconnected portions arranged in a semicircle. One of these portions was about 200 feet wide and 35 feet high. This moraine was mainly a ground moraine, probably forced up by some elevation of the ledge under the ice.

Another border moraine to the north of Kornok's northern ice river [near latitude 64° 40′] was of a different character. The stones, at least at the surface, were greatly in preponderance over the gravel. They were angular and of varying size. The moraine showed some arcuations, but taken as a whole it was parallel to the land. In some exceptional instances it approached closely to the land, even so as to touch one of the projecting points, but generally it was located some distance away from land. Its width was estimated at 100 feet, and its height at more than 50 feet; it should be remembered, however, that it might have had a core of ice. Its length was about a mile and a half. South of this moraine, and farther in on the ice, were seen three more moraines, the greatest one extending about 1,000 feet in length. Two of them were parallel, one inside the other.

These observations by Holst are obviously analogous with those of Prof. I. C. Russell, in 1890 and 1891, of the superglacial drift on the borders of the Malaspina ice sheet in Alaska, made superglacial by ablation of the ice, and bearing forest trees and luxuriant shrubby and herbaceous vegetation. Again, Holst's notes of the oc· currence of the englacial drift up to definite heights of 200 to 500 feet have been repeated for the glaciers and border of the Greenland ice sheet on the north side of Inglefield Gulf, near latitude 78°, by Prof. T. C. Chamberlin, in 1894, who finds the englacial drift to extend to heights of 50 to 100 feet, and occasionally 150 feet, or about half of the height of the ice cliffs, being succeeded above by pure ice.

Near the front of the ice sheet in southern Greenland, its contour, fluctuations in extent, and physical condition late in summer, are noted by Holst as follows:

The ice within a hundred feet from its borders invariably presents a slope toward the border, though generally not so steep as to render the ascent at all difficult. Farther in the slope is much less marked, though there appears to exist a general rising toward the east, while the surface everywhere presents vast undulations. The border of the ice appeared to have retreated quite recently in many places: in others it had evidently advanced. This seems to be the necessary effect of the varying amount of precipitation of snow or rain over the glacier basin, causing the glacier itself to vary in volume. The snow fallen in winter seems to remain much longer on the inland ice than on the land. . . . On the surface the inland ice either presented the appearance of a compact mass of coarse crystallinic texture, reminding one of the grain of common rock candy, or else it is honeycombed by the solar heat and shows intersecting systems of parallel plates, apparently the remnants of large ice crystals, often several inches long, which have wasted away, only leaving the frame, as it were, on which they were built. These plates or tablets are highly mirroring, reflecting the solar rays in all directions, depending on the position of each individual crystal.

Nordenskjöld's second expedition on the Greenland ice sheet was in 1883, starting July 4th from almost the same place (near latitude 68° 20′) as in 1870, and advancing nearly due east in eighteen days about 73 miles on to the inland ice, to a height of about 4,950 feet. On July 21st the main party, numbering eight, stopped on account of the wetness and softness of the snow; but two Lapps, travelling with the peculiar snowshoes called *ski*, advanced a probable distance of 45 or 50 miles farther, where the barometers indicated a height of 5,850 feet. Land in the interior, free of ice and bearing vegetation, which Nordenskjöld believed to exist, favoured by *foehn* winds (described on page 123) blowing inland over the mountainous borders of the country, and which he hoped to reach, was not found. Indeed, no nunatak, or projecting top of hill or mountain above the ice surface has been yet discovered more than forty or fifty miles inside the ice-covered area.

A detailed narrative of this journey, from which the following extracts are taken, written by Baron Nordenskjöld in letters to Mr. Oscar Dickson, the patron of the expedition, was given in *Nature* for November 1 and 8, 1883, from which it was reprinted in America by *Science* : *

During the entire journey we had great difficulty in finding suitable camping places. Thus, either the ice was so rough that there was not a square large enough for our tent, or else the sur-

---

* Vol. ii (December 7, 1883), pp. 732–738, with map.

face was so covered with cavities—which I will fully describe later on—that it was necessary to pitch it over some hundred smaller and a dozen larger round hollows one to three feet deep, filled with water, or else to raise it on a snowdrift so loose and impregnated with water that one's feet became wet even in the tent. An exception to this was the place where we camped on July 9th —viz., camping place No. 6. We encountered here a small ice plain surrounded by little rivers, and almost free from cavities, some thirty metres square. All the rivers flowed into a small lake near us, the water from which rushed with a loud roar through a short but strong current into an enormous abyss in the ice plateau. The river rushed close to our tent through a deep hollow, the sides of which were formed of magnificent perpendicular banks of ice. I had the spot photographed, but neither picture nor description can give the faintest idea of the impressive scene—viz., a perfectly hewn aqueduct, as if cut by human hands in the finest marble, without flaw or blemish. Even the Lapps and the sailors stood on the bank lost in admiration.

. . . On the 11th . . . we proceeded alongside a big river, the southern bank of which formed a comparatively smooth ice plain, or, rather, ice road, with valleys, hills, cavities, or crevasses. . . . This plain was in several places beautifully coloured with "red" snow, especially along the banks of the river. It was the only spot on the whole inland ice where we found "red" snow or ice in any quantity. Even yellow-brown ice was seen in some places; but, on the other hand, ice coloured grayish-brown or grayish-green, partly by kryoconite and partly by organisms, was so common that it generally gave colour to the ice landscape.

Even on July 12th, between camps Nos. 7 and 8, we found blades of grass, leaves of the dwarf birch, willows, crackberry, and pyrola, with those of other Greenland flora, on the snow. At first we believed that they had been carried hither from the interior, but that this was not the case was demonstrated by the circumstance that none was found east of camp No. 9. The only animals we discovered on the ice were, besides the few birds seen on our return journey, a small worm which lives on the various ice algæ and thus really belongs to the fauna of the inland ice, and two storm-driven birds from the shore. I had particularly requested each man to be on the lookout for stones on the ice; but, after a journey of about half a kilometre from the ice border, no stone

19

was found on the surface, not even one as large as a pin's point. But the quantity of clay-dust (kryoconite) deposited on the ice was very great—I believe several hundred tons per square kilometre.

. . . The 9th camp lay on the west side of an ice ridge close by a small, shallow lake, the water from which gathered, as usual, into a big river, which disappeared in an abyss with azure-coloured sides. From this spot we had a fine view of the country to the west, and saw even the sea shining forth between the lofty peaks on the coast; but when we reached east of this ice ridge the country was seen no more, and the horizon was formed of ice only.

Through an optical illusion, dependent on the mirage of the ice horizon, it appeared to us as if we were proceeding on the bottom of a shallow, saucer-shaped cavity. It was thus impossible to decide whether we walked up or down hill, and this formed a constant source of discussion between us, which could only be decided by the heaviness of the sledges in the harness. . . .

The constant sunshine by day and night, reflected from every object around, soon began to affect our eyes—more so, perhaps, because we had neglected to adopt snow spectacles at the outset of our journey; and snow blindness became manifest, with its attendant cutting pains. Fortunately, Dr. Berlin soon arrested this malady—which has brought so many journeys in the arctic regions to a close—by distributing snow spectacles, and by inoculating a solution of zinc vitriol in the blood-stained eyes. Another malady, if not so dangerous, at all events quite as painful, was caused by the sunshine in the dry, transparent, and thin air on the skin of the face. It produced a vivid redness and a perspiration with large burning blisters, which, shrivelling up, caused the skin of the nose, ears, and cheeks to fall off in large patches. This was repeated several times, and the pain increased by the effect of the cold morning air on the newly formed skin. Any similar effect the sun has not in the tropics. With the exception of these complaints, none of us suffered any illness.

On July 13th we covered thirteen, on the 14th ten, and on the 15th fourteen kilometres (9th to 12th camps). At first the road gradually rose, and we then came to a plain which I, in error, believed was the crest of the inland ice. The aneroids, however, showed that we were still ascending; thus, the 9th camp lies 753, the 10th 877, the 11th 884, and the 12th 965 metres above the sea. Our road was still crossed by swift and strong rivers, but the ice became more smooth, while the kryoconite cavities became more and more troublesome. This was made more unpleasant by rain, which began to fall on the afternoon of July 13th, with a heavy wind from the southeast. It continued all the night, and the next morning turned into a snowstorm. We all got very wet, but consoled ourselves with the thought that the storm coming from the southeast argued well for an ice-free interior. When it cleared a little we strained our eyes to trace any mountains which would break the ice horizon around us, which everywhere was as level as that of the sea. The desire soon " to be *there* " was as fervent as that of the searchers of the El Dorado of yore ; and the sailors and the Lapps had no shadow of doubt as to the existence of an ice-free interior ; and at noon, before reaching camp No. 12, everybody fancied he could distinguish mountains far away to the east. They appeared to remain perfectly stationary as the clouds drifted past them—a sure sign, we thought, of its not being a mass of clouds. They were scanned with telescopes, drawn, discussed, and at last saluted with a ringing cheer; but we soon came to the conclusion that they were unfortunately no mountains, but merely the dark reflection of some lakes farther to the east in the ice desert.

. . . The kryoconite cavities were perhaps more dangerous to our expedition than anything else we were exposed to. . . . These lie, with a diameter just large enough to hold the foot, as close to one another as the stumps of the trees in a felled forest, and it was therefore impossible not to stumble into them at every moment ; which was the more annoying, as it happened just when the foot was stretched for a step forward, and the traveller was precipitated to the ground with his foot fastened in a hole three feet in depth. The worst part of our journey was four days outward and three days of the return ; and it is not too much to say that each one of us, during these seven days, fell a hundred times (daily) into these cavities—viz., for all of us, seven thousand times. I am only surprised that no bones were broken. . . . One advantage the kryoconite cavities had, however, viz., of offering us the purest drinking water imaginable, of which we fully availed ourselves without the least bad consequences, in spite of our perspiring state.

On July 16th we covered thirteen, on the 17th eighteen and a half, and on the 18th seventeen and a half kilometres. The country, or, more correctly, the ice, now gradually rose from 965 to 1,213 metres. The distances enumerated show that the ice became more smooth, but the road was still impeded by the kryoconite cavities, whereas the rivers, which even here were rich in water, became shallower but stronger, thus easier of crossing. Our road was, besides, often cut off by immense snow-covered crevasses, which, however, did not cause us much trouble.

. . . During these days we passed several lakes, some of which had the appearance of not flowing away in the winter, as we found here large ice blocks several feet in diameter, screwed up on the shore ; which circumstance I could only explain by assuming that a large quantity of water still remained here when the pools about became covered with new ice. The lakes are mostly circular, and their shores formed a snow " bog," which was almost impassable with the heavy sledges.

On July 19th we covered seventeen and a half, on the 20th sixteen and a half, on the 21st seven, and on the 22d seven and a half kilometres (15th to 18th camps). The ice rose between them from 1,213 to 1,492 metres. The distances enumerated fully show the nature of the ice. It was at first excellent, particularly in the morning, when the new snow was covered with a layer of hard ice ; but on the latter days we had great difficulty in proceeding, as a sleet fell with a southeast wind in the night between the 20th and 21st. The new snow, as well as that lying from the previous year, became a perfect snow bog, in which the sledges constantly stuck, so that it required at times four men to get them out. We all got wet, and had great difficulty in finding a spot on the ice dry enough to pitch the tent. On the 22d we had to pitch it in the wet snow, where the feet immediately became saturated on putting them outside the India-rubber mattresses. A little later on in the year, when the surface of the snow is again covered with ice, or earlier, before the thaw sets in, the surface would no doubt be excellent to journey on. . . . It being utterly impossible to get the sledges farther, I had no choice. I decided to turn back.

I wished, however, to let the Lapps go forward some distance to the east to see the country as far as possible. . . . At 2.30 A. M. on July 22d they started. The days we waited for them were generally spent in the tent, as water surrounded us everywhere. The sky was covered with a thin veil of clouds, through which the sun shone warmly, at times even scorchingly. From time to time this veil of clouds, or haze, descended to the surface of the ice and hid the view over the expanse; but it was, remarkably enough, not wet, but *dry*—yes, so dry that our wet clothes absolutely dried in it. . . .

On the 24th, after an absence of fifty-seven hours. the Lapps returned. . . . As to the run, Lars rendered the following report: When they had reached thirty miles from the camp, no more water could be found. Farther on, the ice became perfectly smooth. The thermometer registered $-5°$ C. It was very easy to proceed on the *skidor*. At the point of return the snow was level, and packed by the wind. There was no trace of land. They only saw before them a smooth ice covered by fine and hard snow. The composition of the surface was this: first, four feet of loose snow, then granular ice, and at last an open space large enough to hold an outstretched hand. It was surrounded by angular bits of ice (crystals). The inland ice was formed in terraces, thus: first a hill, then a level, again another hill, and so on. The Lapps had slept for four hours, from twelve, midnight, on July 23d, in a hollow dug in the snow, while a terrific storm blew. They had till then been awake fifty-three hours. . . . On the return journey . . . two ravens were seen; they came from the north, and returned in the same direction. . . .

On July 25th we began the return journey. It was high time, as the weather now became very bad, and it was with great difficulty we proceeded in the hazy air between the number of crevasses. The cold, after the sun sank below the horizon at night, also became very great, and on the morning of July 27th the glass fell to $-11°$ C. The rivers now impeded us but little, as they were to a great extent dried up. The ice-knolls had decreased considerably in size, too, and lay more apart; but the glacial crevasses had greatly expanded, and were more dangerous, being covered with snow. Even the cavities and the glacial wells, of which many undoubtedly leave a veritable testimony of their existence behind them in the shape of corresponding hollows in the rock beneath, had expanded and increased in number. On a few occasions on the return journey we saw flocks of birds, most probably waterfowl, which were returning from the north.

Robert E. Peary, in June and July, 1886, accompanied by Christian Maigaard, made the next important exploration of the inland ice, going east from the head of Pakitsok Fiord on the northeast part of Disco Bay, in latitude 69° 30'. An account of this expedition, which was a reconnoissance, with the hope of the more extended journey made six years later, is given by Peary in the Bulletin of the American Geographical Society, New York.*

The explorers advanced to a distance of about 100 miles from the edge of the ice, attaining an altitude of about 7,500 feet. Describing the first ten miles of the ice, Peary writes: " In detail, the surface was, as a rule, roughly granular in texture, affording firm, sure footing, interrupted here and there by crevasses, some open, and some covered with a snow arch by patches of soft, deep snow in the depressions between the hummocks, and by patches of hard ice cut by nearly parallel furrows, as if made by a huge plough." The camp at the end of their advance was in a shallow basin of the névé of snow which covers all the inner portion of the ice sheet, there having, to use Peary's words, " the consistency of fine granulated sugar as far down as I could force my alpenstock (some six feet)." The margin and the interior of the ice sheet are characterized by Peary as follows:

Wherever the ice projects down a valley in a long tongue or stream the edges contract and shrink away from the warmer rocks on each side, leaving a deep cañon between, usually occupied by a glacier stream. . . . Higher up along the unbroken portions of the dam [i. e., inclosing mountains], where the rocks have a southern exposure or rise much above the ice, there is apt to be a deep cañon between the ice and the rocks, the ice-face

---

* Vol. xix (September 30, 1887), pp. 261-289.

sometimes 60 feet high, pure pale green, and flinty. In another place the ice-face may be so striated and discoloured as to be a precise counterpart of the rock opposite, looking as if torn from it by some convulsion. The bottom of the cañon is almost invariably occupied by water. . . . Still farther up, at the very crest of the dam, the ice lies smoothly against the rocks.

As to the features of the interior beyond the coast line, the surface of the ice blink near the margin is a succession of rounded hummocks, steepest and highest on their landward sides, which are sometimes precipitous. Farther in, these hummocks merge into long, flat swells, which in turn decrease in height toward the interior, until at last a flat, gently rising plain is reached, which doubtless becomes ultimately level.

In concluding the narrative of this journey, after describing the needful outfit, Peary remarked : " To a small party thus equipped, and possessed of the right mettle, the deep, dry, unchanging snow of the interior . . . is an imperial highway, over which a direct course can be taken to the east coast." It was also suggested that the unexplored northern shore lines of Greenland may be most readily mapped by expeditions across the high inland ice.

Two years later, in August and September, 1888, Dr. Fridtjof Nansen, with five companions, crossed this ice sheet from east to west between latitude 64° 10' and 64° 45', being the first to learn the contour and slopes of an entire profile over the inland ice. The width of the ice there is about 275 miles, extending into the ocean on the east, but terminating on the west about 14 miles from the head of Ameralik Fiord, and 70 miles from the outer coast line. For the first 15 miles in the ascent from the east, rising to the altitude of 1,000 metres, or 3,280 feet, the average gradient was nearly 220 feet per mile. In the next 35 miles an altitude of 2,000 metres, or 6,560 feet, was reached; and

the average gradient in this distance, between 15 and
50 miles from the margin of the ice, was thus about 94
feet per mile, or a slope very slightly exceeding one
degree. The highest part of the ice sheet, about 112
miles from the point of starting, was found to have an
altitude of 2,718 metres, or about 8,920 feet. Its ascending slope, therefore, in the distance from 50 to 112
miles, was about 38 feet per mile. Thence descending
westward, the gradients are less steep, averaging about
25 feet per mile for nearly 100 miles to the altitude of
2,000 metres, about 63 feet per mile for the next 52
miles of distance and 1,000 metres of descent, and
about 125 feet per mile for the lower western border of
the ice.

The narrative of this expedition is most admirably
told by Dr. Nansen in two finely-illustrated volumes, entitled The First Crossing of Greenland. The scientific
results attained are presented in an appendix of the
second volume, from which the following extracts are
quoted :

As to the superficial aspect of the inland ice, I may say, in the
first place, that of crevasses we found a surprisingly small number in the course of our journey. On the east side they occurred
only in the first seven or eight miles; on the west side we came
across the first fissure at some twenty-five miles from the margin
of the ice. In the interior there was no trace of them.

Of surface rivers we found practically none. Some may be
inclined to think that this was due to the lateness of the season,
though this objection has little force, seeing that the middle of
August, when we were on the east side, is not late in the season
as far as regards the melting of the snow; and, furthermore, that
even if the rivers had disappeared themselves on the west coast,
we should have seen traces of their channels. None such did we
see in the interior at all, and the first we observed were not more
than fifteen or twenty miles distant from the western edge. It is
possible, also, that there were minor brooks on the surface in the

first ten miles from the eastern side. Except for these small water courses near the two coasts, I may say positively that there are no rivers at any time of the year on that part of the inland ice over which we passed.

. . . At no great distance from the east coast the surface of dry snow begins, on which the sun has no other effect than to form a thin crust of ice. The whole of the surface of the interior is precisely the same. . . .

Of moraine *débris* or erratic blocks we met with none upon the ice, with the exception of the last little slope when we left it for good on the western side, or no more than a hundred yards from the extreme edge. . . .

. . . Some of the temperatures which we experienced were far lower than the established meteorological laws could have led us to expect. . . . The temperature on certain nights, September 12th and 14th, probably fell, according to the calculations of Professor Mohn, to −45° C. (−49° F.), while the mean temperature of certain days, September 11th to 16th, when we were about in the middle of the country, or a little to the west of the highest ridge, varied from −30° C. to −34° C. (−22° to −29° F.). This is at least 20 C. (36° F.) lower than any one would have been justified in expecting, if he had based his calculations on accepted laws, taking for his data elevation above and distance from the sea, as well as the mean temperature of the neighbouring coasts.

. . . In the forty days which we spent on the ice there were sixteen of either snow or rain. On four days only did we have rain, when we were weather-bound in the tent near the east coast, and on one day near the west coast we had hail; on the rest it was always snow, which in the interior took the form of fine " frost snow," or needles of ice. This fell almost daily out of a half-transparent mist, through which we could often see the sun, together with halos and mock suns.

The severe temperatures experienced by Nansen and his party were in remarkable contrast with the prevailingly warm weather and abundant snow-melting which Nordenskjöld encountered somewhat earlier in the summer five years previously, at a distance of only about three hundred miles farther north. The diversity of average character of seasons in different years, which is often observed in temperate latitudes of the United States and Europe, appears also to be equally exhibited in Greenland. We can not doubt that during the middle three or four weeks of the summer in 1883 the surface of the great *névé* covering the Greenland ice sheet was rapidly thawing, with many resulting super-glacial streams, upon the area traversed by Nansen; but such warm and fast melting weather seems probably to be exceptional, occurring perhaps only once in several years, like the times of severe drought which rarely come, as one in ten or twenty years, more or less, to the eastern United States, or like our occasional prolonged thaws in the middle of the winters.

The most extensive journey of exploration thus far accomplished on the Greenland ice sheet was by Lieutenant Robert E. Peary and Eivind Astrup in 1892. The narrative of this expedition, sailing from New York, June 6, 1891, wintering on the south shore of McCormick Bay, at the northern side of the entrance of Inglefield Gulf, near latitude 77° 40', and performing a sledge journey of about thirteen hundred miles, including both the advance and return, on the inland ice of north-western Greenland, has been well told by Peary and others of his party, in the Bulletin of the American Geographical Society,[*] in a volume entitled In Arctic Seas,[†] and especially in My Arctic Journal, by Mrs.

* Vol. xxiii, pp. 444–454, September 30, 1891 ; vol. xxiv, pp. 470–473 and 536–558 (with maps and views from photographs), September 30 and December 31, 1892.

† In Arctic Seas : The Voyage of the Kite with the Peary Expedition, by Robert N. Keely, Jr., M. D., and G. G. Davis, A. M., M. D. ; vii, 524 pages, 1892.

Peary,[*] who accompanied her husband to McCormick
Bay and spent a year there.

Peary, with three comrades, began the journey on
the ice sheet May 3d; but the next two weeks were
employed in transporting their supplies up the steep
and crevassed ice border, and over the rough upward
slope for the first fifteen miles, to the beginning of the
vast interior snow plain.   During this time, and again
within the following week, they were hindered by severe
snowstorms, with "the constant violent wind rushing
down from the interior to the shore," such as Hayes
had experienced there in 1860; but from the 24th of
May, when two of the party returned to the station at
McCormick Bay, no other violent storms occurred during the northward march.   Taking a northeastward
course past the heads of the depressions leading to the
Humboldt Glacier, Petermann Fiord, and the Sherard
Osborne Fiord, Peary and his young Norwegian companion, Astrup, reached the vicinity of the northern
end of the ice sheet June 27th, where, near latitude 82°
and longitude 40°, they saw mountainous but lower
land and a fiord in front of them, on account of which
they changed their course to the east and southeast.
On July 1st they set out across the land, and after
"four days of the hardest travelling, over sharp stones
of all sizes, through drifts of snow and across rushing
torrents," reached a headland which was named Navy
Cliff, whence they looked down about thirty-five hundred
feet upon Independence Bay, so named from its dis-

---

[*] My Arctic Journal: A Year among Icefields and Eskimos,
by Josephine Diebitsch-Peary: with an Account of The Great
White Journey across Greenland, by Robert E. Peary, Civil Engineer, United States Navy: 240 pages, 1893.

covery on July 4th. A large glacier, discharging from
the inland ice, flows from the south into this bay, near
latitude 81° 37′ and longitude 34°. Peary's description
of this northeastern border of Greenland, and of his
return journey, is as follows:

> This land, red and brown in colour, and almost entirely free
> of snow, is covered with glacial *débris* and sharp stones of all
> sizes. Flowers, insects, and musk oxen are abundant. We shot
> five musk oxen and a large number of birds. Traces of foxes,
> hares, ptarmigan, and possibly wolves, were seen. The surface of
> the bay was covered with winter's still unbroken ice, prisoning
> the icebergs from the great glacier.
>
> On July 9th we started on the return, taking a course more in-
> land. In seven days we were struggling through the soft snow,
> and wrapped in the snow-clouds of the great interior plateau,
> over eight thousand feet above the sea level. We remained in
> the clouds some fourteen days, when we descended from them
> east of the Humboldt Glacier. Then, with dogs and ourselves
> trained down to hard pan, we covered over thirty miles per day
> for seven days, till our eyes were gladdened by the deep green,
> iceberg-dotted waters of McCormick Bay.
>
> On the last day, as I came over the summit of the great ice
> dome lying between the border of the true inland ice and the
> head of the bay, I saw moving figures a mile or two ahead, on the
> next ice dome. From that party burst almost instantly a cheer,
> and it was not long before I was clasping hands with Prof. Heil-
> prin and his men, who were out on a reconnoissance preparatory
> to going in toward Humboldt Glacier to meet me.*

The relief expedition sent out by the Philadelphia
Academy of Natural Sciences, was brought by the same
steamer, the Kite, which had carried Peary to Greenland
the year before, and was under the direction, as before,
of Prof. Angelo Heilprin.† Returning southward, the

---

* Bulletin, American Geographical Society, vol. xxiv, pp. 472,
473.

† The Arctic Problem, and Narrative of the Peary Relief Ex-

Kite, bearing Peary and his party, left their station, called Redcliffe House, August 24th, and reached the Waigat passage, north of Disco Island, August 29th; St. John's, Newfoundland, September 11th; and Philadelphia, September 24th.

Describing the ice sheet whose northwestern part had been thus explored, Peary writes:

The terms "inland ice" and "great interior frozen sea," two of the more common names by which the region traversed by us is generally known, both suggest to the majority of people erroneous ideas. In the first place, the surface is not ice, but merely a compacted snow. The term "sea" is also a misnomer, in so far as it suggests the idea of a sometime expanse of water subsequently frozen over. The only justification for the term is the unbroken and apparently infinite horizon which bounds the vision of the traveller upon its surface. Elevated as the entire region is to a height of from four thousand to nine thousand feet above the sea level, the towering mountains of the coast, which would be visible to the sailor at a distance of sixty to eighty miles, disappear beneath the landward convexity of the ice cap by the time the traveller has penetrated fifteen or twenty miles into the interior, and then he may travel for days and weeks with no break whatever in the continuity of the sharp, steel-blue line of the horizon.

The sea has its days of towering, angry waves, of laughing, glistening white-caps, of mirrorlike calm. The "frozen sea" is always the same—motionless, petrified. Around its white shield the sun circles for months in succession, never hiding his face except in storms. Once a month the pale full moon climbs above the opposite horizon, and circles with him for eight or ten days.

Sometimes, though rarely, cloud shadows drift across the white expanse, but usually the cloud phenomena are the heavy prophecies or actualities of furious storms veiling the entire sky; at other times they are merely the shadows of dainty, transparent cirrus feathers. In clearest weather the solitary traveller upon this white Sahara sees but three things outside of and beyond

---

pedition of the Academy of Natural Sciences of Philadelphia, by Angelo Heilprin; 165 pages (Contemporary Publishing Co., 1893).

himself—the unbroken white expanse of the snow, the unbroken
blue expanse of the sky, and the sun. In cloudy weather all three
of these may disappear.

Many a time I have found myself in cloudy weather travelling
in gray space. Not only was there no object to be seen, but in
the entire sphere of vision there was no difference in intensity of
light. My feet and snowshoes were sharp and clear as silhouettes,
and I was sensible of contact with the snow at every step; yet,
as far as my eyes gave me evidence to the contrary, I was walking
upon nothing. The space between my snowshoes was as light as
the zenith. The opaque light which filled the sphere of vision
might come from below as well as above. A curious mental as
well as physical strain resulted from this blindness with wide-
open eyes, and sometimes we were obliged to stop and await a
change.

The wind is always blowing on the great ice cap, sometimes
with greater, sometimes with less violence, but the air is never
quiet. When the velocity of the wind increases beyond a certain
point it scoops up the loose snow, and the surface of the inland
ice disappears beneath a hissing white torrent of blinding drift.
The thickness of this drift may be anywhere from six inches to
thirty or even fifty feet, dependent upon the consistency of the
snow. When the depth of the drift is not in excess of the height
of the knee, its surface is as tangible and almost as sharply defined as that of a sheet of water, and its incessant dizzy rush and
strident sibilation become, when long continued, as maddening
as the drop, drop, drop, of water on the head in the old torture
rooms.

. . . As a result of my study of the Eskimo clothing and its
use, I adopted it. . . . The deerskin coat, with the trousers, foot-
gear, and undershirt, weighed eleven and one fourth pounds, or
about the same as an ordinary winter business suit, including
shoes, underwear, etc., but not the overcoat. In this costume,
with the fur inside and the drawstrings at waist, wrists, knees, and
face pulled tight, I have seated myself upon the great ice cap,
four thousand feet above the sea, with the thermometer at −38°,
the wind blowing so that I could scarcely stand against it, and
with back to the wind have eaten my lunch leisurely and in comfort; then, stretching myself at full length for a few moments,
have listened to the fierce hiss of the snow driving past me with

the same pleasurable sensation that, seated beside the glowing grate, we listen to the roar of the rain upon the roof.

Our sleeping-bags, also of the winter coat of the deer, with the fur inside, were, I think, the lightest and warmest ever used. In my own bag, weighing ten and one fourth pounds, I have slept comfortably out upon the open snow, with no shelter whatever, and the thermometer at −41 , wearing inside the bag only under-garments. During the inland-ice journey, throughout which the temperature was never more than a degree or two below zero, our sleeping-bags were discarded, our fur clothing being ample pro-tection for us when asleep, even though I carried no tent.*

In the summer of 1893, Lieutenant Garde, of the Danish Navy, made an expedition upon the inland ice, starting from the glacier of Sermilsialik, latitude 61°, on the eastern side of Greenland. The distance travelled was 300 kilometres (186 miles), occupying thirteen days, and the highest elevation attained was 7,000 feet. The marches were at night, as the *névé* was then in better condition than during the warmer daytime.†

Again in 1893, Lieutenant and Mrs. Peary, with a party for further exploration of the ice sheet and north-ern shores of Greenland, sailed in the Falcon from Port-land, Me., July 8th, and from St. John's on the 14th, reaching Bowdoin Bay, on the north side of Inglefield Gulf, August 3d. The narrative of this voyage to their new winter station, named Anniversary Lodge, at the head of Bowdoin Bay, near latitude 77° 40', longitude 70°, is interestingly told by Mrs. Peary in the closing chapter of My Arctic Journal.

---

* The Great White Journey, My Arctic Journal, pp. 231–233, 239, 240.

† Revue de Géographie, September, 1893, cited by Bulletin of the American Geographical Society. vol. xxv, p. 439, September 30, 1893.

From this station, very early in the spring of 1894, on the 6th day of March, Peary started, with the plan of travelling northeast over the Greenland ice sheet a distance of about 650 miles to Independence Bay, the limit of his previous expedition, thence intending to send one party south, while he, with one or more assistants, would explore the country farther north. On setting out, the party comprised eight men, twelve sledges, and ninety dogs. The time, however, proved to be much too early, on account of the severity of weather on the high ice sheet at the very beginning of the circumpolar half year of constant daylight. After a journey of two weeks on the inland ice, reaching an altitude of about 5,000 feet, the party experienced, on March 20th to the 23d, an "equinoctial storm" of blinding snow, fierce wind, and very low temperature, probably unequalled in the experience of any former arctic expedition. The self-recording anemometer showed that the wind during thirty-four hours had an average velocity of forty-eight miles an hour; and the thermograph showed an average temperature of 50° F. below zero. Exceedingly cold weather and other severe storms followed, the temperature being mostly 40° to 50° below zero, with almost continual wind. Some of the men had their feet and hands frozen; the dogs, enduring in the snow outside the tents the full hardships of the storms, were in a few instances frozen to death, and the others were attacked by a fatal disease:* and some of the sledges were broken in being drawn over the sharply ridged snowdrifts. The party was soon diminished to half its original number by the return of frost-bitten and sick men, until the expedition, after having ad-

---

* Called *piblockto*, before noticed on page 240.

vanced in total about 125 miles, was reluctantly abandoned by Peary on April 10th, that he might save a
sufficient reserve of his provisions, sledges, and dogs for
another attempt the next year. The summer was spent
in explorations of the Greenland coast, glaciers, and border of the ice sheet, in the neighbourhood of the winter
station and south to Melville Bay.*

In August, 1894, the Falcon again came from Newfoundland to Inglefield Gulf, as a relief expedition to
bring back Peary and his party; but the intrepid explorer, with two comrades, remained in Greenland for
another year, while Mrs. Peary and the others of the
party returned. It was Peary's hope, during the summer of 1895, to succeed in crossing the ice sheet and to
explore more fully the northeastern and northern coast.
The very severe storms encountered in the early spring
indicate that travel on either the ice sheet of Greenland
or that of the antarctic continent, where a distance of
850 miles lies between the most southern indentation of
the shore line and the pole, will be practicable only during a few months in the middle and later parts of the
circumpolar summers.

The geologist of the Peary relief party of 1894, sailing in the Falcon, was Prof. T. C. Chamberlin, whose
principal purpose was to study the plane of contact of
the ice sheet and land surface, the drainage of the wast-
ing ice margin, and all possible phases of glacial deposition. This party, under the command of Henry G.
Bryant, Secretary of the Geographical Club of Philadelphia, and well known for his exploration of the Grand
River in Labrador, left Brooklyn, N. Y., June 20th, and

* Bulletin of the American Geographical Society, vol. xxvi, pp.
397–406, September 30, 1894.

St. John's July 7th.   Greenland was first sighted July
12th, and its coast was followed northward more than
1,000 miles to Peary's winter station on Inglefield Gulf.
A landing was made on Disco Island July 16th, where
three glaciers were examined.   This island was also
visited on the return of the expedition, allowing the
same glaciers to be seen again forty-eight days later.
Much ice was encountered in Melville Bay, so that the
Falcon was unable to make the middle passage, but the
inner passage was traversed without serious delay, reaching Cape York July 23d.   The next day was given to a
search on the Carey Islands for further information regarding the lost Swedish naturalists, Björling and Kal-
stenius, wrecked there two years before, of whose party
numerous relics were found.   Prof. Chamberlin writes:

On the morning of July 25th the Falcon entered Whale Sound,
the mouth of Inglefield Gulf, our destination, but found it covered
with ice still too strong to permit the forcing of a passage.   This
was the day fixed for our arrival in the prospectus of the expedition, and had the gulf been open we would have reached Peary's
headquarters on schedule time.   The trip up to this time had been
nearly ideal from the standpoint of one who wished to see the
realities of the arctic region, without suffering much from them.
We had some sharp battling with the ice pack, some groping in
the fog, were beset and nipped with moving floes, but were not
very seriously threatened or long delayed.   We saw just enough
of the vicissitudes of the region to realize what they might become in their full force, and just enough of the dangers to give
us a wholesome respect for them.

The vessel being unable at once to reach Peary's headquarters,
work was begun upon the glaciers immediately at hand, and
dredging was commenced with excellent results.   Communication
was soon established over the gulf ice with headquarters, and
Lieutenant Peary and several of his party visited the Falcon.   On
August 5th I returned with Lieutenant Peary to his headquarters
by his invitation, and remained his guest until nearly the time for
our return.   Meanwhile the Falcon and the rest of the auxiliary

party, and several of the Peary party, went to Ellesmere Land
and Jones Sound for geographic study and for dredging. They
found the ice unusually extensive and solid, and their results were
necessarily limited. On returning, the Falcon forced her way
with difficulty through the ice to the Peary headquarters, arriving August 20th. On the 26th the return was begun. A short
call was made at Cape York, and stoppages of two days each at
Godhavn and Godthaab, the two capitals of Greenland. St.
John's was reached without notable incident on September 15th,
and Philadelphia on the 25th.

Peary's headquarters are surrounded by glaciers, some of which
are tongues of the great inland ice, while some come from local
ice caps; some reach the sea level and give birth to icebergs, while
some terminate inland. Some have gentle declines and deploy on
open ground, while others have steep gradients and crowd through
narrow valleys. The great inland ice cap is less than three miles
distant. The facilities for glacial study are unsurpassed.*

In his address as President of the Geological Society
of America, at its Baltimore meeting, December 28,
1894, Prof. Chamberlin gave a summary of his observations on the glaciers and ice sheet in the Inglefield Gulf
region, of which the following is an abstract, from notes
taken during the address:

The glaciers in the vicinity of Bowdoin Bay, where terminating on the land, commonly have very steep, often nearly vertical
and sometimes overhanging fronts to heights of one hundred to two
hundred feet or more. This remarkable contrast with the glaciers
of more southern latitudes is ascribed to peripheral melting by
reflection of the oblique solar rays from the warm adjoining
ground. It seems also to be secondarily dependent on the very
slow rate of the glacial motion, which is found by Peary to be
usually scarcely measurable, while the maximum daily rate observed in exceptionally fast-flowing glaciers in the midsummer is
from two and a half to four feet.

Englacial drift is plentifully seen in many of the frontal ice

---

* Glacialists' Magazine, vol. ii. pp. 69, 70, November, 1894.

cliffs to heights of fifty to one hundred feet, and occasionally a hundred and fifty feet, or about half of their total height. It is quite unequally distributed, being commonly gathered, especially

FIG. 51.—East face of Bryant Glacier, Inglefield Gulf. Showing vertical wall and stratification of the ice. (Chamberlin.)

at considerable heights, into layers of an inch to a foot or more, where the ice contains much rock detritus, interbedded with thicker layers of nearly pure ice. Again, masses of drift several feet in extent, analogous with till, are rarely enveloped in the ice, which, above and beneath these masses and the similarly inclosed boulders, has an upwardly and downwardly arching lamination.

Differential onward flow of the ice, its upper and middle portions outstripping those next beneath, has been the chief means of giving to the terminal ice-cliff sections a very distinctly and surprisingly laminated structure. Sometimes the differential movement has carried part of a previously plane zone forward so much faster than that which was before it as to bend the clearly laminated zone into sigmoid folds, and even to produce sharply defined overthrust faults. In this way the englacial boulders and small rock fragments are frequently much worn and striated.

The inclusion of the englacial drift is attributed to abrasion from knobs and ridges projecting up into the overriding ice, rather than to any upwardly flowing basal currents. It is observed, however, that the front of the ice sometimes rides upward over its own marginal accumulations of drift; indeed, wherever onward movement of the ice boundary is taking place this seems more frequent than any glacial erosion or pushing forward of the moraine. Strong winds blow prevailingly down the slope of the inland ice, often drifting snow over the frontal glacier cliffs, and even these snowdrifts are sufficient in some places to cause the ice to flow upward, or to be laterally deflected.

In some cases the morainic hillocks, seen in process of formation beneath the steep or vertical edge of the ice, have the outlines of miniature drumlins, with the laminated glacier curving upward quite conformably over them. No eskers or kames were observed. The drainage from the glacial melting is mostly by subaërial lateral streams, along the inner side of the adjoining moraines ; rarely it is by central subglacial streams.

Only very scanty drift is spread over the country outside the ice sheet and glaciers; and the largest glacio-fluvial delta fans are about half a mile in extent. Most of the glaciers have been long stationary: a few are retreating; others are advancing.

Near the east side of Bowdoin Bay a driftless area, having a diameter of three or four miles, shows deep decomposition of its rock, which is hornblendic gneiss. Its altitude is less than that of neighbouring glaciers, and it is accepted, with the jagged and unglaciated outlines of the upper part of many of the coastal mountains from Capes Farewell and Desolation north to Inglefield Gulf (a distance of twelve hundred miles), as decisive evidence that there has never been a complete envelopment of the western border of Greenland by land ice.

Dalrymple Island, close to the Greenland coast, near longitude

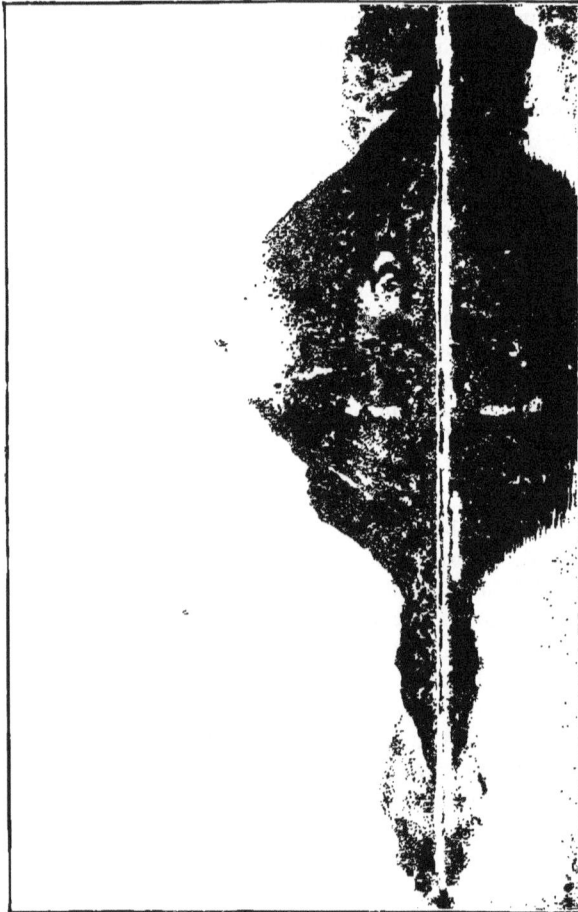

FIG. 52.—Dalrymple Island, near the Greenland Coast. Showing unglaciated profile. (Chamberlin.)

70° and latitude 76° 30′, also consists of decomposing hornblendic gneiss, with no drift, and with mountain forms due solely to sub-aërial erosion. Fifty miles northwestward, however, the Carey

Islands, which are mountains rising from a large expanse of the surrounding northern part of Baffin Bay, have been glaciated by an ice sheet flowing over them from the north—that is, from Grinnell Land and Smith Sound. In the course of this ice sheet, at a distance of fifty miles north of the Carey Islands, the sea has a depth of two hundred and twenty fathoms.

Inquiring for the physical causes and explanation of glacial motion, Prof. Chamberlin thinks the theory of Hugi, Grad, Forel, and others, which refers the movement, under the influences of the solar heat and gravity, to the enlargement and long persistence of the granules originating in the *névé*, to be more supported by his observations and studies than the now commonly accepted theory of J. D. Forbes, which regards the ice as a viscously flowing though brittle solid.*

In the present year, 1895, with his two companions, Peary traversed the Greenland ice sheet upon nearly the same course as three years before, from Anniversary Lodge, at the head of Bowdoin Bay, which was left April 1st, to Independence Bay. Musk oxen were found there, as in the previous expedition, and several were shot, affording an important addition to the scanty provisions of the party, since, with only one exception, the caches made at the end of the attempted journey in March and April, 1894, had become buried in the snow and could not be found. The caches had indeed been searched for in vain by an expedition during the intervening September, for the purpose of raising them from the summer snow and again marking their locations.

---

* The American Geologist, vol. xv. pp. 197. 198. March, 1893. This address is published in the Bulletin of the Geological Society of America, vol. vi, pp. 199–220, with eight plates. Upon a wider range than could be given in this address, the same studies of the Greenland glaciers and ice sheet are being presented very fully by Prof. Chamberlin in a series of articles in the Journal of Geology, from the number for October-November, 1894, onward.

Great hardships were encountered in this journey, and
of the forty-one dogs, drawing three sledges, with which

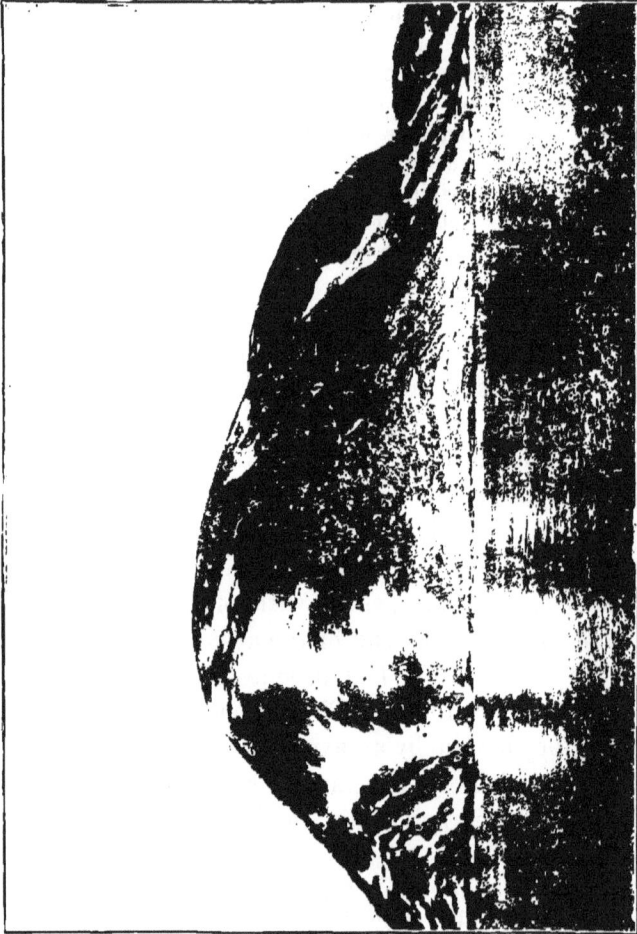

Fig. 58.—Southeastern Carey Island. Showing characteristic glaciated contour. (Chamberlin.)

the advance from the vicinity of the lost caches began,
only eleven survived to reach the northern border of the
ice sheet. Peary's hoped-for opportunity to explore the

northeastern coast in the neighbourhood of Independence Bay was partially attained during a prolonged hunt for musk oxen; but the exhaustion of one of his comrades, and the scarcity of their food, which would evidently be wholly needed for the homeward journey, compelled them soon to turn back.

The return was begun with nine dogs and ended with only one, reaching Anniversary Lodge June 25th. For more than two weeks the three men had subsisted on one meal a day, and they had no food during the last march of twenty-one miles.

The relief expedition of this year, in the steamer Kite, though passing unimpeded through the usually ice-covered Melville Bay, was unable on account of floe ice to enter Bowdoin Bay, but reached McCormick Bay, separated from the preceding by the Redcliff Peninsula. Members of the relief party, walking to Anniversary Lodge, found Peary and his comrades on August 3d, and the homeward voyage to St. John's, Newfoundland, was completed September 21st.

Prof. R. D. Salisbury in this relief expedition, as Prof. Chamberlin in that of the previous year, made extensive observations of the ice cap and glaciers of Redcliff Peninsula and of the western coast of Greenland, thence southward to Disco Island and the Jakobshavn glacier. Lieutenant Peary also, in geographic exploration of the Greenland coast from Cape York northward 150 miles in latitude to Cape Alexander, had accurately mapped about a hundred glaciers, and had taken a long series of observations of the rate of motion of one of the most active glaciers in the Inglefield Gulf region.

It should be added that the Kite brought back from Greenland a very valuable zoölogical collection, which was made principally by Prof. L. L. Dyche, including about 4,000 specimens of birds and their eggs, narwhals, seals, polar bears, and other animals. Lastly, two large masses of iron, one weighing about three tons, were obtained from the coast of Melville Bay. These are thought to be meteorites, but it seems worthy of inquiry whether their origin may be like that of the equally large iron masses erupted from the earth's interior with basaltic lava, which by decay and weathering leaves the iron masses on the surface, at Ovifak, much farther south in western Greenland.*

---

* Lieutenant Peary's narrative of his difficult and perilous journey on the ice sheet this year, with notes of his other explorations and of the collections by this relief party, is in the Bulletin of the American Geographical Society, vol. xxvii, pp. 300–306, September 30, 1895.

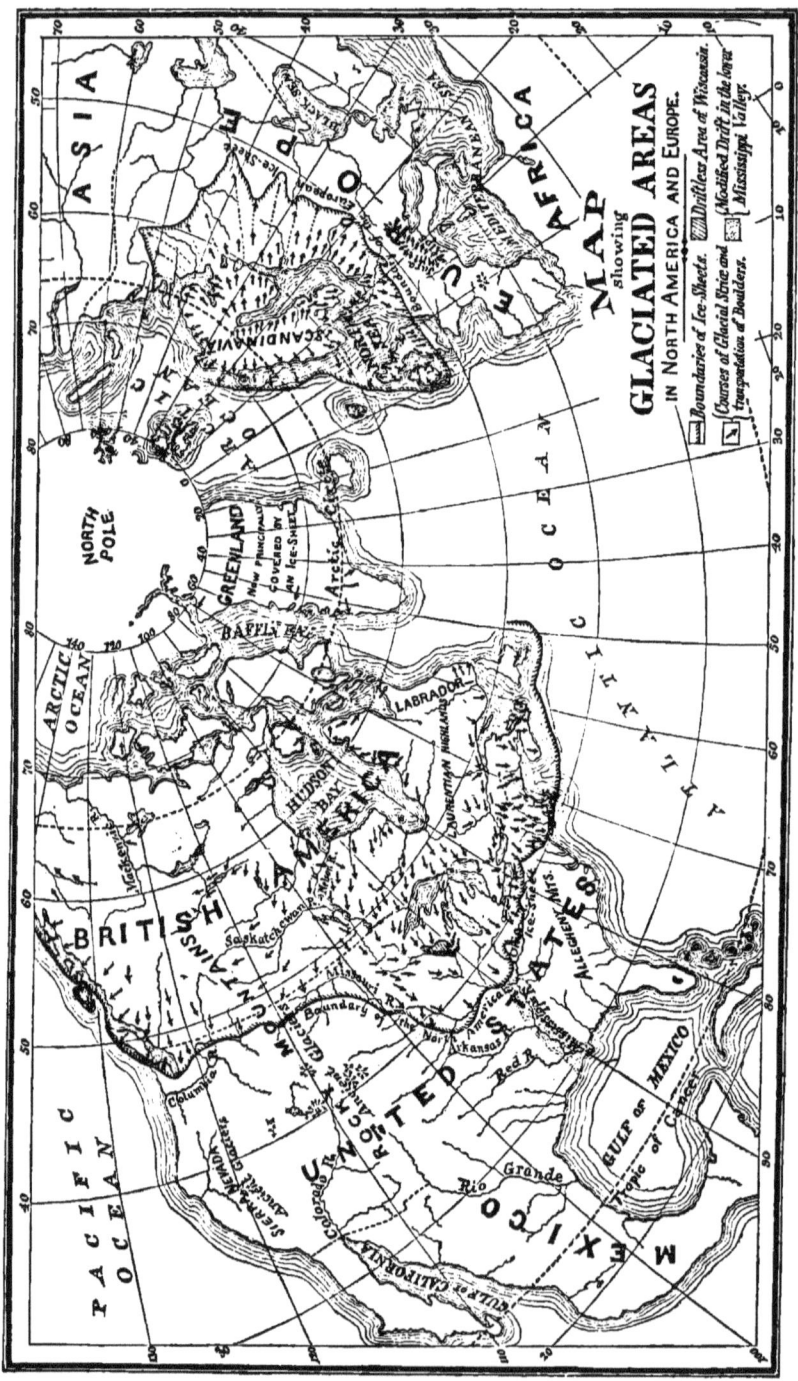

MAP
showing
GLACIATED AREAS
IN NORTH AMERICA AND EUROPE.

Boundaries of Ice-Sheets. Driftless Area of Wisconsin.
Course of Glacial Stria and Modified Drift in the Iowa
transportation of Boulders. Mississippi Valley.

CHAPTER XI.

THE guiding principle of geologic investigation, brought out most clearly by Sir Charles Lyell, requires us to seek the explanation of past changes of the earth by observation and study of agencies which are now in operation, producing similar changes during the present epoch. Such studies of the Swiss glaciers by Agassiz, Forbes, Tyndall, and others, proved conclusively, a generation ago, that the drift was formed by land ice, so that the comparatively small district of the Alps supplied the clew for deciphering the records of the latest completed chapter of the geologic history of northwestern Europe and the northern half of North America. Glaciers of other regions in the Eastern hemisphere, notably of the Himalayas and of Norway, have also contributed much to our knowledge of the ice sheets of the Pleistocene or Glacial period. The vast ice sheets of that time, however, are adequately exemplified at the present day only by the antarctic and Greenland ice sheets, less completely and on a much smaller scale by the yet very instructive Malaspina Glacier, and in some respects they may be profitably compared with the Muir Glacier, which is the most fully studied icefield of America or perhaps of the world.

In seeking to derive from the descriptions of the

297

icefields of Greenland, as given in the foregoing pages, their full significance for the explanation of the methods of formation of the glacial and modified drift deposits in temperate latitudes, we may therefore well notice briefly these other now existing ice sheets and glaciers.

Land ice surrounds the south pole to a distance of twelve to twenty-five degrees from it, covering, as Sir Wyville Thomson estimates, about 4,500,000 square miles. Its area is thus slightly greater than that of the Pleistocene ice sheet of North America, which covered about 4,000,000 square miles; while the confluent Scandinavian and British ice sheets appear to have enveloped no more than 2,000,000 square miles, including the White, Baltic, North, and Irish Seas, whose areas were then occupied by the continental *mer de glace*. The observations of Ross, in sailing along a part of the border of the antarctic ice sheet, and of Moseley, in the cruise of the Challenger among the enormous tabular icebergs which float away from it, have been noted in the preceding chapter (page 246).

The Malaspina ice sheet in Alaska, stretching from the St. Elias range to the shore of the Pacific Ocean, has been described as follows by its principal explorer, Prof. I. C. Russell, after his two expeditions of 1890 and 1891 : *

This glacier extends with unbroken continuity from Yakutat Bay 70 miles westward, and has an average breadth of between

* Mount St. Elias and its Glaciers, American Journal of Science, III, vol. xliii. pp. 169–182, with map, March, 1892. The report of the first expedition. in 1890, is given by Russell in the National Geographic Magazine, vol. iii. pp. 53–203, with 19 plates and 8 figures in the text, May 29, 1891.

20 and 25 miles; its area is approximately 1,500 square miles, . . . a vast, nearly horizontal plateau of ice, with a general elevation of about 1,500 feet. The central portion is free from moraines and dirt, but is rough, and broken by thousands and tens of thousands of small crevasses. Its surface is broadly undulating, and recalls the appearance of portions of the rolling prairie lands west of the Mississippi. . . . On looking down on the glacier from an elevation of 2,000 or 3,000 feet on the hills bordering it on the north, even on the wonderfully clear days that follow storms, its limits are beyond the reach of vision. From any commanding station overlooking the Malaspina glacier, as from the summit of the Chaix Hills, for example, one sees that the great central area of clear, white ice is bordered on the south by a broad, dark band formed of boulders and stones. Outside of this, and forming a belt concentric with it, is a forest-covered area, in many places four or five miles wide. . . .

The moraines not only cover all of the outer border of the glacier, but stream off from the mountain spurs that project into its northern border. . . . The stones and dirt previously contained in the glacier are . . . concentrated at the surface, owing to the melting of the ice that contains them. This is the history of all of the moraines of the Malaspina glacier. They are formed of the *débris* brought out of the mountains by the tributary alpine glaciers, and concentrated at the surface by reason of the ablation of the ice. . . .

The outer and consequently older portions of the fringing moraines are covered with vegetation, which in places, particularly near the outer margin of the belt, has all the characteristics of old forests. It consists principally of spruce trees, some of which are three feet in diameter, and cottonwood, alder, and a great variety of shrubs and bushes, together with rank ferns, which grow so densely that one can scarcely force a passage through them. The vegetation grows on the moraines resting on the ice, which in many places is not less than a thousand feet thick. . . . It is only on the stagnant border of the ice sheet that forests occur. The forest-covered area is, by estimate, between twenty and twenty-five square miles in extent.

The drainage of the Malaspina glacier is almost entirely interglacial or subglacial. There is no surface drainage, excepting in a few localities where there is a surface slope, but even in such places the streams are short, and soon plunge into a crevasse or a moulin and join the drainage beneath.

On the lower portions of the alpine glaciers, tributary to the Malaspina, there are sometimes small streams coursing along in ice channels, but they are short-lived. On the borders of these tributaries there are frequently important streams flowing between the ice and a mountain slope, but where these come down to the Malaspina they flow into tunnels and are lost to view.

Along the southern portion of the Malaspina glacier, between the Yahtse and Point Manby, there are hundreds of streams which pour out of the escarpment formed by the border of the glacier, or rise like great fountains from the gravel and boulders at its base. All of these streams are brown and heavy with sediment, and overloaded with boulders and stones.

One of the largest streams draining the glacier is the Yahtse. This rises in two principal branches at the base of the Chaix Hills, and, flowing through a tunnel some six or eight miles long, emerges at the southern border of the glacier as a swift, brown flood, fully one hundred feet across and fifteen or twenty feet deep. The stream, after its subglacial course, spreads out into many branches, and has built up an alluvial fan which has invaded and buried thousands of acres of forest. In traversing the coast from the Yahtse to Yakutat Bay we crossed scores of ice-water streams which drain the icefield to the north. The greater part of these could be waded, but some of them are rivers which it was impossible to ford.

Orogenic (mountain-building) disturbances formed the St. Elias range since the beginning of the Pleistocene period, for its basal portion consists of the late Pliocene and Pleistocene Pinnacle and Yakutat formations, above which the St. Elias schist has been overthrust. Fossil marine shells, all of which are represented by species now living in the adjacent ocean, were collected by Russell in the cliffs above Pinnacle Pass, at the height of five thousand feet above the sea. The following summary of the history of this range is given by Russell : [*]

* National Geographic Magazine, vol. iii, pp. 172, 173.

Not only was a part, at least, of the Pinnacle system deposited during the life of living species of molluscs, but also the whole of the Yakutat series, the stratigraphic position of which is, if my determination is correct, above the Pinnacle system. After the sediments composing the rocks of these two series were deposited in the sea as strata of sand, mud, etc., they were consolidated, overthrust, faulted, and upheaved into one of the grandest mountain ridges on the continent. Then, after the mountains had reached a considerable height, if not their full growth, the snows of winter fell upon them, and glaciers were born. The glaciers increased to a maximum, and their surface reached from a thousand to two thousand feet higher than now on the more southern mountain spurs, and afterward slowly wasted away to their present dimensions. All of this interesting and varied history has been enacted during the life of existing species of plants and animals.

The Muir Glacier, which was explored in 1886 by Wright* and Baldwin,† and in 1890 by Reid‡ and Cushing,# is situated some 200 miles southeast of Mount St. Elias, and about 125 miles north of Sitka. It has an estimated area, with its tributary glaciers, of 350 square miles, and the area inclosed by its watershed is about 800 square miles. The slope of the main glacier for 10 miles or more next to its termination in the sea at the head of Glacier Bay is about 100 feet per mile. Its frontal cliffs, which shed multitudes of bergs into the sea, had in 1890 an extent of 1¾ miles, and rose in a vertical wall of ice 130 to 210 feet above the water, which within 300 feet from the ice front has a maximum

---

* American Journal of Science, III, vol. xxxiii, pp. 1–18, with map, January, 1887 ; The Ice Age in North America, 1889, chapter iii.

† American Geologist, vol. xi, pp. 366–375, June, 1893.

‡ National Geographic Magazine, vol. iv, pp. 19–84, with 16 plates and 5 figures in the text, March 21, 1892.

American Geologist, vol. viii, pp. 207–230, with map, October, 1891.

depth of 720 feet. Between the years 1886 and 1890 the front had receded one third to two thirds of a mile.

In 1886 the height of the Muir ice cliffs above the water was found to be 250 to 300 feet; and the rates of forward motion of the most prominent ice pinnacles near to the front and within half a mile back from it were roughly measured and found to vary from 72 feet to 9 feet per day, the maximum being that of a pinnacle close to the projecting middle of the terminal cliffs. In 1890 the rates of the glacial currents were measured by observations of flags set on the surface of the glacier one fourth to one half of a mile back from its then nearly straight and much lower front, and the maximum movement at some distance from the centre was ascertained to be only about seven feet per day. In respect to the apparent discrepancy of these determinations, separated by an interval of four years, it is to be remarked that the ice pinnacles, belonging to the most fractured and crevassed portions of the glacier, doubtless move onward much faster than its more even tracts, which can be traversed and marked by flags; and that the two different years between which the front was withdrawn so far may have been considerably unlike in the meteorologic conditions governing the flow of the glacier. The abundant observations of Helland, Steenstrup, Drygalski, and others, on the rates of outflow of glaciers into the fiords and bays of the western coast of Greenland, show that there the glacial advance ranges frequently from 30 to 65 feet daily, and in at least one case is about 100 feet. Narrow glaciers in the valleys of the Alps move only a few feet daily; but the broad glaciers of polar regions, when they terminate in the sea, often move at their ends much more rapidly, as 30 to 50 feet or more per day.

Like the Malaspina ice sheet and other glaciers of the St. Elias region, the Muir Glacier is fast retreating. From the narrative of Vancouver's exploration of this coast in 1794, and from observations of freshly glaciated rock surfaces far outside and also far above the present glacier, it appears sure that only one to two centuries ago the Muir Glacier stretched some 20 miles farther than now, nearly to the mouth of Glacier Bay. Its advance to this maximum area had perhaps occupied a considerably longer time than its retreat, but the whole time of both advance and recession appears to be geologically recent. In its forward movement forests became enveloped in the gravel and sand discharged by streams from the glacier, and they were then overridden by the ice advance, so that now on its retreat the still standing trees are being uncovered by the channelling of streams.

During the summer of 1890 the rate of ablation of the frontal part of the Muir Glacier was about 14 inches per week, which would lower its surface probably 15 or 20 feet in the whole season. This corresponds approximately with the ablation of the Mer de Glace in Switzerland, ascertained by Forbes to be 24½ feet between June and September, in 1842; while in some exceptional cases the ablation of glaciers in summer has been found to be as much as one foot a day.

One other point of great significance brought out by these investigations of the Muir Glacier is the approximate determination of the rate of glacial erosion upon its rock bed. Measurements of the sediment in the water of the copious subglacial streams, and estimates of the quantity of water annually discharged from the rainfall and snowfall of the Muir basin, indicate,

21

according to Wright's computations, an average erosion of one third of an inch yearly for all the ice-clad
area.

From these existing glaciers and ice sheets we learn
much concerning the probable surface slopes and thickness of the ice sheets of the Glacial period. In North
America, the upper limits of the Pleistocene glaciation
on Mount Katahdin, the Catskills, the Three Buttes or
Sweet Grass Hills of Montana, the Rocky Mountains
north of the international boundary, and the mountains
of British Columbia give us reliable information of the
thickness of the ice sheet in the vicinity of these high
elevations of land. Its depth is computed by Dana to
have been about two miles on the Laurentide highlands,
between the St. Lawrence and Hudson Bay, whence the
ice flowed radially outward in all directions, during its
maximum stage overtopping the White, Green, and
Adirondack Mountains. In British Columbia, according to Dr. G. M. Dawson, its maximum depth was about
7,000 feet. These thicknesses, however, which seem
well determined, would not give to the borders of the
North American ice sheet surface slopes of more than
about 25 to 30 feet per mile; whereas the Greenland ice
sheet is known to have surface gradients of 100 to 200
feet per mile. Apparently, slopes of at least 50 feet or
more per mile for the outer portion of the ice sheet
are required to produce strong glacial currents, such as
transported boulders 1,000 miles, from the eastern side
of the southern part of Hudson Bay, where it narrows
into James Bay, southwestward to southern Minnesota,
and such as carried Scandinavian boulders likewise
about 1,000 miles from their sources to central Russia.
The Pleistocene ice sheets could have had gradients
comparable in steepness with those of the Greenland

ice sheet only by widely extended uplifts of the central
portions of their areas to heights at least several thousand feet above their present altitudes. These continental movements of uplift, suggested by the ice sheet
of Greenland, appear, as will be shown in a later chapter, to have been the cause of the climatic changes by
which the ice sheets of North America and Europe
were accumulated.

Another comparison may be made in respect to the
rates of erosion by the Pleistocene ice sheets. The
Muir Glacier is found to be eroding its rock bed at the
rate of about a third of an inch in a year, or a foot
in thirty-six years, and nearly three feet in a century.
Probably the erosion by the ice sheets of the Glacial
period was less rapid, even in the zone of most efficient
action, which may have been usually from 50 to 200
miles inside the ice boundary. If their erosion was a
half or third as much, in a thousand years this zone
would be eroded to an average depth of 10 feet, more
or less, and in five thousand years 50 feet. Farther
within the ice-covered area the erosion was probably
less; but during the recession of the ice sheet, and perhaps also during its accumulation, the maximum rate of
erosion would prevail successively upon all parts of the
drift-bearing regions. In the light of this comparison
with the modern Muir Glacier, it is evident, from the
volume of the drift and the topographic features of the
country, that a geologic period not exceeding thirty
thousand or fifty thousand years would suffice for the
observed volume of the Pleistocene glacial erosion and
resulting drift.

With reference to the questions whether the drift
was transported chiefly beneath the ice sheets of the
Glacial period or chiefly within their lower part, the

observations of Holst and Chamberlin on the englacial * drift of the Greenland ice sheet and of Russell on the Malaspina Glacier fully reaffirm the conclusions reached through investigations of the North American drift by Dana, Shaler, C. H. Hitchcock, N. H. Winchell, Upham, and others, that the Pleistocene ice sheet contained much drift in its lower portion to heights of probably 1,000 to 1,500 feet above the ground. This transportation of drift within the ice seems to have been equally important with the subglacial transportation, and indeed probably to have much exceeded that in its amount.

The rates of observed ablation of the Muir Glacier and the Mer de Glace suggest that during the closing stage of the Glacial period the ice sheets, both in America and Europe, may have been melted away very fast. If such ablation prevailed every summer for one or two centuries, it must melt 2,000 to 4,000 feet of ice, which was approximately the thickness of the Pleistocene ice sheet from central New England westward across the Laurentian lakes to Minnesota and southern Manitoba. This accords with the apparent duration of the glacial Lake Agassiz for only about 1,000 years, and with the evidently very rapid accumulation of the eskers, of their associated sand plains and plateaus, of the valley drift, and of the drumlins. There were, indeed, many times of halt or readvance of the ice front, interrupting its general retreat, as shown by the marginal moraines, of which Chamberlin, Leverett, and others have mapped no less than fifteen or twenty in their order from south to north in Ohio, Indiana, Illinois, and southern Michi-

---

* A term proposed by Prof. Chamberlin to designate drift incorporated in the ice and thus transported with it.

gan; but such halts forming large moraines on each
side of Lake Agassiz were demonstrably of short continuance, only for a few decades of years, and the whole
departure of the ice sheet from the southern end of that
glacial lake to Hudson Bay was geologically very rapid.

The Malaspina ice sheet has been gradually retreating during the past hundred years, or probably much
longer. On all its border for a width of a few miles,
now thinned perhaps to a quarter part, or less, of
the earlier depth, the waning ice is covered by its formerly englacial drift, but in that cold climate the
glacial movement is so very slow that forest trees, with
luxuriant undergrowth of shrubs, and many herbaceous
flowering plants, grow on this drift lying upon hundreds
of feet of ice, as revealed by stream channels. Advancing toward the interior, the explorer soon comes upon
higher clear ice and *névé*, having risen above the plane
of the englacial *débris*, excepting along the course of
belts of medial surface morainic drift, swept outward
from spurs of the mountains. This ice sheet partially
suggests the conditions of the moraine-forming southern
portion of the North American and European ice sheets
during the Champlain epoch or closing part of the Ice
Age; but these had a climate much warmer than that
of Alaska, with consequently far more rapid ablation and
stronger glacial currents.

In Greenland, on the other hand, the mean temperature has probably been gradually lowered during several centuries past, since the prosperous times of the
Norse colonies, nine hundred to five hundred years ago.
A great ice sheet 1,500 miles long, with a maximum width
of 700 miles and an area of about 575,000 square miles,
covers all the interior of Greenland; and although now
its extent is less than during the Glacial period, it has

doubtless held its own, or mainly somewhat increased during several hundred years. While the snow and ice accumulation is predominant, no englacial drift becomes superglacial; but in the region of Inglefield Gulf Chamberlin finds the frontal ice cliffs well charged with

Fig. 54.—Gable Glacier, Inglefield Gulf, showing inset débris and lamination of the ice. (Chamberlin.)

englacial *débris* to a third or half of their total heights of 100 to 200 feet or more. The same ratio of the lower part of the ice sheet containing drift would quite certainly give it a thickness of 1,000 to 2,000 feet in the deeply ice-covered central portion of Greenland. Other features especially noted are the very distinct stratification of the ice and its differential forward motion, producing not only this stratification but also sigmoid folds and overthrust faults, where the upper layers move faster than the lower, and these in turn faster than the friction-hindered base. In just the same way the accelerated currents of the waning ice sheet during the temperate Champlain epoch overrode each other in succession from the highest to the lowest on the moraine-forming border, bearing a great amount of superglacial drift to the margin, whenever it remained nearly stationary during a series of years. If a mild temperate climate could bring to Greenland the conditions of the Champlain epoch, its thick ice sheet in the interior under rapid ablation would fully illustrate, as the Malaspina Glacier even now does in a considerable degree, the formation of the great series of morainic drift hills which mark stages in the retreat of the continental ice sheets.

CHAPTER XII.

In North America, on the west side of the North Atlantic Ocean, and in Europe, on its east side, ice sheets of vast extent were accumulated during the Ice Age or Pleistocene period—the last completed division of geologic time—immediately preceding the present or Psychozoic period. Between the two great areas that were then ice enveloped is the land, almost a continent, which has been the theme of the foregoing pages; and we have seen that it now lies, for the greater part of its whole area, beneath an ice sheet similar to those of North America and Europe in Pleistocene times. Greenland, therefore, may be well expected to supply us the key for the interpretation of the processes of the erosion, transportation, and deposition of the glacial drift on the contiguous continents. Furthermore, in our inquiry concerning the causes of the extraordinary Pleistocene ice accumulation, we may well look to Greenland for aid toward the solution of this difficult and still much debated question. Reserving these discussions, however, to later chapters, it will be profitable here to take a general survey of the changes in the relative heights of land and sea which are known to have occurred on the northern, western, and eastern borders of the North Atlantic basin during the Pleistocene

310

period. These land oscillations not only account for the peculiarities of geographic range of many species in the flora and fauna of Greenland, but also they have

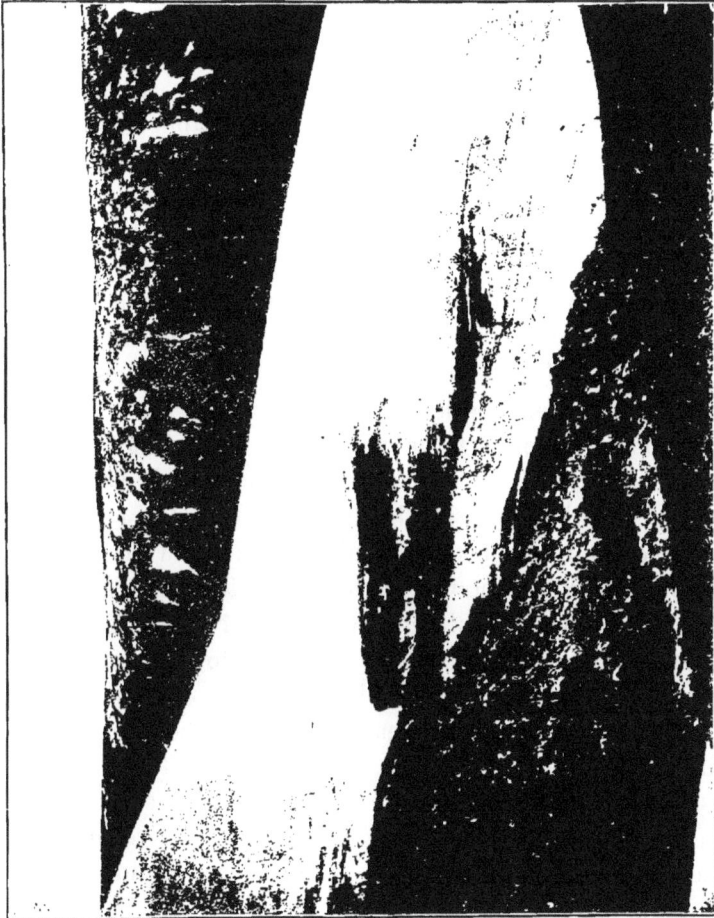

Fig. 55.— North side of Gable Glacier, showing inthrust of layers of débris. (Chamberlin.)

given origin to the great climatic changes of the arctic and north temperate zones, through which the warm temperate Miocene circumpolar flora was succeeded at the end of the Tertiary era by the ice sheets adjoining the North Atlantic, and later, by returning warmth, melting away the continental glaciers and somewhat restricting the area of the great field of snow and ice enveloping the interior of Greenland. In the present chapter we will examine the evidences of the Pleistocene changes of level, leaving their application as the causes of the accumulation, and later of the disappearance of the Pleistocene ice sheets, for special review in another chapter.

More than forty years ago Prof. James D. Dana showed that the fiords are valleys which were eroded by streams during a formerly greater elevation of the land. At the time of their excavation the streams flowed along the bottom of the fiords, and the depths of these beneath the present sea level are a measure of the ensuing subsidence, or of a large part of it, for evidently the subsidence could not have been less, but may have been more, than the depth of the fiords. Few of the fiords of Greenland have been carefully sounded, but many of the ice fiords are known, by the size of the bergs floated out to sea with the ebbing of the tides (10 to 12 feet in difference of height between high and low tide on the Baffin Bay coast), to have at least a depth of 1,000 feet of water. In the Jakobshavn ice fiord the fishermen sink their lines to a depth of nearly 1,300 feet for the fish which have their feeding grounds at the bottom. In the Franz Josef Fiord, on the east side of Greenland, Captain Koldewey found no bottom by a sounding of 500 fathoms (3,000 feet).

Greenland is divided from the contiguous North American continent and archipelago by a great valley of erosion, which is estimated from soundings and tidal records to have a mean depth of 2,510 feet below sea level for 680 miles through Davis Strait, 2,095 feet for 770 miles next northward through Baffin Bay, and 1,663 feet for the next 55 miles north through Smith Strait.* Continuing northward, this eroded valley, now depressed beneath the sea, has a depth of 203 fathoms (or 1,218 feet) in Kennedy Channel, latitude 81°.

The fiords of the west coast of Greenland vary from 10 or 20 miles to 75 miles or more in length ; and on the east coast the deep and branching Franz Josef Fiord (latitude 73° 15′) was explored by Koldewey in 1870 to a distance of about 100 miles from the outer shore line.

Analogous in their physical features and origin with the fiords are the long, irregular straits, channels, and sounds which divide the many large and small islands of the arctic archipelago west of Baffin Bay. They are doubtless old river courses of the former northward continuation of the continent, now so far submerged that what were valleys have become branching and interlocking arms of the sea. Only scanty soundings have been obtained in these waters, which are covered by floes and the broken ice pack during the greater part of the year. In Lancaster Sound, much frequented by American whalers, the depth of 900 feet is recorded ; in Barrow Strait, leading thence westward between North Devon and the west part of Cockburn Land, there is a sounding of 1,680 feet ; and the north part of Prince Regent's Inlet, leading southward near 90° longitude, has a depth of 1,080 feet. (United States Navy Hydrographic Office charts.)

---

* Smithsonian Contributions to Knowledge, vol. xv, pp. 163, 164. Am. Geologist, vol. vi, p. 330, December, 1890; Am. Jour. Sci., III, vol. xlvi, p. 119, August, 1893. (Compare also the statement of the same explanation in a more recent paper, by Prof. N. S. Shaler, Bulletin, Geol. Soc. Am., vol. vi, p. 158, January, 1895.)

Southward, on the northeastern borders of North America, the British Admiralty and United States Coast Survey charts, according to Prof. J. W. Spencer, who has given much attention to these oscillations of altitude, record submerged fiord outlets from Hudson Bay,

Fig. 56.—North side of Gable Glacier, Inglefield Gulf, showing an overthrust, with débris along the plane of contact. The ice is much veined. (Chamberlin.)

the Gulf of St. Lawrence, and the Gulf of Maine, respectively 2,040 feet, 3,666 feet, and 2.664 feet below sea level. The bed of the old Laurentian River from the outer boundary of the Fishing Banks to the mouth of the Saguenay, a distance of more than 800 miles, is reached by soundings 1,878 to 1,104 feet in depth. Advancing inland, the sublime Saguenay Fiord, along an extent of about 50 miles, ranges from 300 to 840 feet in depth below the sea level, while in some places its bordering cliffs, 1 to 1½ mile apart, rise abruptly 1,500 feet above the water.*

The continuation of the Hudson River Valley has been traced by detailed hydrographic surveys to the edge of the steep continental slope at a distance of about 105 miles southeastward from Sandy Hook. Its outermost 25 miles are a submarine fiord 3 miles wide and from 900 to 2,250 feet in the vertical depth, measured from the crests of its banks, which, with the adjoining flat area, decline from 300 to 600 feet below the present sea level. The deepest sounding in this submerged fiord is 2,844 feet.†

An unfinished survey by soundings off the mouth of Delaware Bay finds a similar valley submerged nearly 1,200 feet, but not yet traced to the margin of the continental plateau.

Farther to the south Prof. N. S. Shaler concludes that Florida has been uplifted probably 2,000 feet, more or less, above its present height. This opinion is founded on the distribution of species of animals and

---

* J. W. Spencer, Bulletin, Geol. Soc. Am., vol. i, 1890, pp. 65–70, with map of the Gulf of St. Lawrence and the submerged Laurentian River. J. W. Dawson, Notes on the Post-Pliocene Geology of Canada, 1872, p. 41 ; The Canadian Ice Age, 1893, pp. 71–74.

† A. Lindenkohl, Report of U. S. Coast and Geodetic Survey for 1884, pp. 435–438 ; Am. Jour. Sci., III, vol. xxix, pp. 475–480, June, 1885 ; James D. Dana, Am. Jour. Sci., III, vol. xl, pp. 425–437, December, 1890, with an excellent map of the Hudson submarine valley and fiord.

plants in the West Indies and in the adjoining parts of North and South America; on the estuarine mouths of the rivers of the southeastern United States; on the existence of fresh water to the depth of 900 feet, and of salt water lower, in a deep well bored at St. Augustine; and on the occurrence of a great fresh-water spring welling up strongly in the sea "a few miles to the south of St. Augustine and three or four miles from the coast line." The issuance of this submarine spring can be only from a cavernous subterranean water course in the limestone of the Floridian peninsula, originally due to running water when the land was formerly much elevated.* It is worthy of note that an uplift of 2,064 feet would unite this peninsula with Cuba, closing the Strait of Florida, through which the Gulf Stream now pours into the North Atlantic.

In the West Indies the recent geologic researches of A. J. Jukes-Browne and J. B. Harrison,† R. T. Hill,‡ and J. W. Spencer,# well demonstrate that these islands have undergone great changes of level, both of subsidence and uplift, during late Tertiary and Quaternary time. Similar conclusions are also reached by C. T. Simpson from his study of the distribution and variations of the land and fresh-water molluscs of the West Indian region.‖

The Isthmus of Panama also appears to have been depressed deeply beneath the sea within the same late

---

* Nature and Man in America, 1891. pp. 99–107; Bulletin, Geol. Soc. Am., vol. vi, 1895, pp. 154, 155.

† Quart. Jour. Geol. Soc., vol. xlvii, 1891, pp. 197–250.

‡ Am. Jour. Sci., III, vol. xlviii, pp. 196–212, September, 1894.

Bulletin, Geol. Soc. Am., vol. vi, pp. 103–140, with map and sections, January, 1895.

‖ Proc., U. S. National Museum, vol. xvii, 1894, pp. 423–450.

geologic period, as is indicated by the close relationship
of the Pacific and West Indian deep-sea faunas on the
opposite sides of the isthmus, made known through
dredging by Alexander Agassiz.* This testimony, indeed, with that of Darwin, L. and A. Agassiz, and
others, of very recent, extensive, and deep subsidence of
the western coast of South America, apparently, however, continuing for no long time, lends much probability to the supposition that the low Panama isthmus
was somewhat deeply submerged for a geologically short
period contemporaneous with epeirogenic uplifts † of
the circumpolar parts of this continent, both at the
north and south, whereby the effects of great altitude
in covering the northern and southern high areas with
ice sheets were augmented by the passage of much of
the Gulf Stream into the Pacific Ocean.

Several low passes between the Gulf of Mexico and
the Pacific are found in the Lake Nicaragua region, on
the Isthmus of Panama, and in the Atrato River district
on the south, at heights from 133 to 300 feet above the
sea, so that only moderate changes of level would give
a shallow submergence. Previous to the observations
by A. Agassiz in dredging, which seem to require deep
subsidence of the isthmus to account for close alliance
or identity of deep-sea species, a less submergence had
been long before claimed by P. P. Carpenter, Wyville

---

* Bulletin, Museum Comp. Zoöl., at Harvard College, vol. xxi,
pp. 185–200, June, 1891.

† The terms *epeirogeny* and *epeirogenic* (continent-producing,
from the Greek *epeiros*, a mainland or continent) are proposed by
Mr. G. K. Gilbert (in Lake Bonneville, Monograph I, U. S. Geological Survey, 1890, p. 340) to designate the broad movements
of uplift and subsidence which affect the whole or large portions
of the continental areas and of the oceanic basins,

Thomson, and others. Thus Carpenter, in a report on the mollusca of the opposite sides of the Isthmus of Panama, regarded 35 species as alike in the two oceans; 34 other species so nearly allied that they may prove to be identical; and 41 others separated by only very slight differences.* Thomson arranged side by side 18 species of echinoderms from each sea, which resemble each other so closely as to be hardly distinguishable.† A. R. Wallace affirms that on the opposite sides of the isthmus "no less than 48 species of fishes are identical," which, with the community in the lower orders before noted, he thinks to be sufficient proof of "a connection between the oceans at a recent date." ‡

On the Pacific coast of the United States, Prof. Joseph Le Conte has shown that the islands south of Santa Barbara and Los Angeles, now separated from the mainland and from each other by channels 20 to 30 miles wide and 600 to 1,000 feet deep, were still a part of the mainland during the late Pliocene and early Quaternary periods. # In northern California, Prof. George Davidson, of the United States Coast Survey, reports three submarine valleys about 25, 12, and 6 miles south of Cape Mendocino, sinking respectively to 2,400, 3,120, and 2,700 feet below the sea level, where they cross the 100-fathom line of the marginal plateau. ‖ If the land there were to rise 1,000 feet, these valleys

---

* British Association Report, 1856.

† Depths of the Sea, 1873, pp. 13–15.

‡ Nature, vol. ix, p. 220. Jan. 22, 1874. Compare also an article by Dr. Charles Ricketts in the Geological Magazine, II, vol. ii, 1875, pp. 573–580.

Bulletin of the California Academy of Sciences, vol. ii, 1887, pp. 515–520.

‖ Ibid., vol. ii, pp. 265–268.

would be fiords with sides towering high above the water, but still descending beneath it to great depths.

Farther to the north, Puget Sound and the series of sheltered channels and sounds through which the steamboat passage is made to Glacier Bay, Alaska, are submerged valleys of erosion, now filled by the sea, but separated from the open ocean by thousands of islands, the continuation of the Coast Range of mountains. From the depths of the channels and fiords Dr. G. M. Dawson concludes that this area had a preglacial elevation at least about 900 feet above the present sea level, during part or the whole of the Pliocene period.*

Le Conte has correlated the great epeirogenic uplifts of North America, known by these deeply submerged valleys on both the eastern and western coasts, with the latest time of orogenic disturbance by faulting and upheaval of the Sierra Nevada and Coast Ranges in California, during the closing stage of the Tertiary and the early part of the Quaternary era, culminating in the Glacial period.† In the Mississippi basin, from the evidence of river currents much stronger than now, transporting Archæan pebbles from near the sources of the Mississippi to the shore of the Gulf of Mexico, Prof. E. W. Hilgard thinks that the preglacial uplift, inaugurating the Ice Age, was 4,000 or 5,000 feet more in the central part of the continent than at the river's mouth.‡

---

* Canadian Naturalist, new series, vol. viii, pp. 241-248, April, 1877.

† Bulletin, Geol. Soc. Am., vol. ii, 1891, pp. 323-330; Elements of Geology, third edition, 1891, pp. 562-569, 589.

‡ Am. Jour. Sci., III, vol. xliii, pp. 389-402, May, 1892.

The general absence of Pliocene formations along both the Atlantic and Pacific shores of North America indicates, as pointed out by Prof. C. H. Hitchcock, that during this long geologic period all of the continent north of the Gulf of Mexico held a greater altitude, which from the evidence of the submarine valleys is known to have culminated in an elevation at least 3,000 feet higher than that of the present time. Such plateau-like uplift of the continent had only a short duration on the latitude of New York and of Cape Mendocino (between 40° and 41°). With the steep gradients of the preglacial Hudson River and of the streams which formed the now submerged channels on the Californian coast, these rivers, if allowed a long time for erosion, must have formed even longer and broader valleys than the still very impressive troughs which are now found on these submarine continental slopes. But the dura-tion of the epeirogenic uplift of these areas, on the bor-der of the glaciation for the Hudson and beyond it for the Californian rivers, can scarcely be compared in its brevity with the prolonged high altitude held during late Tertiary and early Quaternary time by the deeply fiord-indented Scandinavian peninsula, and by all the northern coasts of North America, from Maine and Puget Sound to the arctic archipelago and Greenland. The abundant long and branching fiords of these north-ern regions, and the wide and deep channels dividing their islands, attest a very long time of preglacial high elevation there. At the time of culmination of the long-continued and slowly increasing uplifts at the north, or, very probably, much later, when the most northern lands and the basin of Baffin Bay had become depressed nearly as now, the maximum epeirogenic elevation seems to have extended during a short epoch far to the south, coincident with the accumulation of ice sheets in high latitudes.

The time of the erosion of the great fiords of Green-land is known to have been wholly subsequent to the deposition of the leaf-bearing beds, the highest of which are considered by Heer to be of Miocene age, while Sir William Dawson and J. Starkie Gardner refer them to the Eocene period. Intercalated with the higher por-tions of these plant-bearing strata of sandstone and shales are layers of basalt, and great outflows of the same lava cap the fossiliferous series (see Chapter VIII, page 208). After the volcanic activity had ceased, an uplift of the land appears to have slowly borne the basaltic plain thousands of feet above its original level; and streams meanwhile, slowly cutting down their beds, channelled the majestic fiords and the deep Waigat Valley whose bottom is now submerged, dividing Disco Island from the Noursoak (or Nugsuak) Peninsula. Of this Waigat passage Dr. Nansen writes: "As Prof. Helland has observed, no geologist who has examined the spot can doubt for a moment that the huge strata from the Cretaceous and Tertiary periods, as well as the later basaltic formations lying above them, which show corresponding features and dimensions on either side of the channel, once formed a solid and connected whole. The channel can not be older than the rocks which form its sides. And as some of the Tertiary deposits are of very recent date, belonging, in fact, to the latest section of Miocene times, and as, furthermore, these strata are covered by huge layers of basalt, it follows that the Waigat itself must be a production of a comparatively recent period. . . . The basalt layers reach a height of from 4,000 to 5,000 feet, or even more; and as the channel itself is of great depth, we can see that here a

mass of solid rock, nearly 100 miles in length, some 10
miles broad, and several thousand feet in thickness, has
been quarried out and carried away. . . ."* Nansen
attributes this stupendous work of erosion to glacial
action when the inland ice was so extended as to fill the
Waigat Valley with an immense outflowing glacier; and
it is true that the ice sheet has been sufficiently greater
than now to permit this assumption. Dr. Rink notes
the occurrence of granitic drift boulders on the top of
the basalt mountains of Disco. The ice has doubtless
occupied the Waigat Valley, but the excavation of this
fiordlike passage, with essentially its present topographic features, like the erosion of the Colorado cañon
and its tributaries, must be ascribed not to ice abrasion,
but to slow, long-continued stream channelling during
a million years, more or less, forming the later part of
the Tertiary era, with some portion probably for the
Greenland fiords, and the whole for the Grand Cañon,
of the ensuing comparatively very short Quaternary era.

For the sake of clearness of statement, we have thus
far considered only or chiefly the great uplifts of the
land areas which characterized the end of the Tertiary
era and the early part of the Pleistocene period up to the
Ice Age. Succeeding that time of general elevation of
the northern and circumpolar lands, on both sides and
north of the North Atlantic Ocean, there came a time
of correspondingly general land depression, when these
countries sank nearly to their present height, or mostly,
for the regions which had been ice-covered, somewhat
lower than now, so that their coastal tracts were to a
considerable extent submerged beneath the sea. The
prevailing connection of this depression with the previ-

---

* The First Crossing of Greenland. vol. ii, pp. 473, 474.

ous heavy loading of the land by the thick mass of the
ice sheets, and their sequence as apparently cause and
effect, will be discussed further on; but it is evident that
in some areas, and especially in the region of northern
Labrador, along Baffin Bay, and farther north, the downward movement of the land, subsequent to the time of
elevation in which the fiords were eroded, was far greater
than would be proportionate with the probable burden
of the ice sheets and their removal.

The greatest amount of this recent subsidence, followed by more recent re-elevation, known for any part
of the great region inclosing the North Atlantic basin,
is in northwestern Greenland and the contiguous Grinnell Land and Grant Land. In a paper read before the
section of Geography in the British Association at the
meeting in August, 1894, Col. H. W. Feilden, who was
the naturalist of the Nares Arctic Expedition in 1875-
'76, reviewed the reasons for hoping and expecting that
Nansen and his party, after drifting in the ice pack
across the sea surrounding the north pole, will return
to tell their experiences. Not only is Siberian driftwood strewn along the northwest coast of Greenland
and the shores of Grinnell Land, but also the currents
have brought such driftwood during a long time past,
in which Grinnell Land has been uplifted at least 1,000
feet. Up to that height Col. Feilden there found
driftwood embedded in recent alluvial or glacial clay
and mud deposits, with marine shells of the species
now living in the adjoining sea. The bivalve shells are
often still held together by their hinges and retain their
brown epidermis; and the wood is combustible, and so
light as to float on water. All the wood appears to be
of coniferous species, being wholly different from the
driftwood cast on the shores of Spitzbergen, which is

borne by the Gulf Stream into the North Atlantic and Arctic Oceans. In another paper, prepared soon after the return of the Nares expedition, Feilden, with his associate, C. E. De Rance, wrote:

> A careful examination of the fossil remains found in the recent beds of Grinnell Land and North Greenland, extending from an altitude of 1,000 feet to the present sea level, gives unmistakable evidence that the fauna is practically identical with that now existing in Grinnell Land as well as in the neighbouring sea. The remains of mammalia, such as the lemming (*Myodes torquatus*), the ringed seal (*Phoca hispida*), the reindeer (*Cervus tarandus*), and musk ox (*Ovibos moschatus*), were discovered in these beds. The marine mollusca most abundant as living species in the adjacent seas, such as *Pecten grœnlandicus*, *Astarte borealis*, *Mya truncata*, and *Saxicava rugosa*, are also the most abundant species throughout the mud beds; while the stems of two species of *Laminaria*, which appear to grow in considerable abundance in the Polar Sea, were detected in mud beds, at an elevation of 200 feet, still retaining their peculiar seashore odour. Coniferous driftwood of precisely the same character as that now stranded was found at elevations of several hundred feet, and so little altered by time or climate that it still retained its buoyancy. . . .
>
> The accounts given of trees having been found in similar post-Pliocene beds in the polar regions, under circumstances that would lead to the supposition that such trees had grown *in situ*, are not to be relied on; and no evidences were discovered in the mud beds of Grinnell Land to encourage the opinion that there have been any interglacial periods of increased temperature—at all events, during the long time which must have elapsed while Grinnell Land was rising to an altitude of 1,000 feet above the present sea level.*

A list of fifteen species of fossil lamellibranch shells, nine of gastropod shells, and several species in other groups, collected by Feilden in the Pleistocene beds of Grinnell Land and northern Greenland up to about

---

* Quart. Jour. Geol. Soc., London, vol. xxxiv, 1878, pp. 566, 567.

1,200 feet above the sea, as identified by Dr. J. Gwyn Jeffreys and others, is given by Sir William Dawson in The Canadian Ice Age (pages 269, 270).

Later, in the same region, fossil marine shells have been found at still greater heights, proving a recent subsidence to the depth of 2,000 feet, with subsequent re-elevation. Gen. A. W. Greely's Report of the United States Expedition to Lady Franklin Bay, Grinnell Land, mentions (in volume ii, page 57) the occurrence there of recent fossil shells of *Astarte lactea* up to 1,000 feet, and of *Saxicava arctica*, as provisionally determined, up to 2,000 feet above the sea. At Polaris Bay, on the neighbouring Greenland coast (near latitude 81° 30'), marine shells are reported, by Dr. Emil Bessels, to occur up to the height of 1,600 feet.

Marine fossils in beds overlying the glacial drift prove that the northeastern part of North America stood lower than now in the Champlain epoch—that is, the time of departure of the ice sheet. This depression, which seems to have been produced by the vast weight of the ice, was bounded on the south approximately by a line drawn from near the city of New York northeastward to Boston and onward through Nova Scotia. When the ice sheet was being withdrawn from this region the country south of this line stood somewhat higher than now, as is shown by the channels of streams that flowed away from the melting ice and ran across the modified drift plains which form the southern shores of Long Island, Martha's Vineyard, Nantucket, and Cape Cod. A subsequent depression of the land there, continuing, perhaps, uninterruptedly to the present time, has brought the sea into these old river courses; but north and northwest of this line the land at the time of recession of the ice sheet was lower than now, and the coast and estuaries were more submerged by the sea. Fossiliferous beds of modified drift, supplied from the melting ice sheet and resting on the till, show that the vertical amount of the marine submergence when the ice sheet disappeared was 10 to 25 feet in the vicinity of Boston and northeastward to Cape Ann; about 150 feet in the vicinity of Portsmouth, New Hampshire; from 150 to 300 feet along the coast of Maine and southern New Brunswick; about 40 feet on the northwestern shore of Nova Scotia; thence increasing westward to 200 feet in the Bay of Chaleurs, 375 feet in the St. Lawrence Valley opposite to the Saguenay, and 560 feet at Montreal; 300 to 400 feet, increasing from south to north, along the basin of Lake Champlain; about 275 feet at Ogdensburg, and 450 feet near the city of Ottawa; 300 to 500 feet on the country southwest of James Bay; and in Labrador, little at the south, but increasing northward to 1,500 feet at Nachvak, according to Dr. Robert Bell; while in northern Greenland and Grinnell Land, as before noted, it was from 1,000 to 2,000 feet.

That the land northward from Boston was so much lower while the ice sheet was being melted away is proved by the occurrence of fossil molluscs of far northern range, including *Yoldia* (*Leda*) *arctica* Gray, which is now found living only in arctic seas, where they receive muddy streams from existing glaciers and from the Greenland ice sheet. This species is plentiful in the stratified clays resting on the till in the St. Lawrence Valley and in New Brunswick and Maine, extending southward to Portsmouth, New Hampshire. But it is known that the land was elevated from this depression to about its present height before the sea here became warm, and the southern molluscs, which exist as colonies in the Gulf of St. Lawrence, migrated thither; for these southern species are not included in the extensive lists of the fossil fauna found in the beds overlying the till.

In the St. Lawrence basin these marine deposits reach to the southern end of Lake Champlain, to Ogdensburg, and Brockville, and at least to Pembroke and Allumette Island, in the Ottawa River, about 75 miles above the city of Ottawa. The Isthmus of Chiegnecto, connecting Nova Scotia with New Brunswick, was submerged, and the sea extended 50 to 100 miles up the valleys of the chief rivers of Maine and New Brunswick.

From the Champlain submergence attending the departure of the ice the land was raised somewhat higher than now, and its latest movement from New Jersey to southern Greenland has been a moderate depression. The vertical amount of this postglacial elevation above the present height and of the recent subsidence on all the coast of New Jersey, Long Island, New England, and the eastern provinces of Canada, is known to have ranged from 10 feet to a maximum of at least 80 feet at the head of the Bay of Fundy, as is attested in many places by stumps of forests, rooted where they grew, and by peat beds now submerged by the sea.

At the time of final melting of the ice sheet this region, which before the Ice Age had stood much higher than now, was depressed, and the maximum amount of its subsidence, as shown by marine fossils at Montreal and northwestward to Hudson Bay, was 500 to 600 feet. Subsequently our Atlantic coast has been re-elevated to a height probably 100 feet greater than now; and during the recent epoch its latest oscillation has been again downward, as when it was ice-covered. The rate of depression since the discovery of America has probably

averaged one to two feet in a hundred years along the
distance from New Jersey to Nova Scotia. At the
ruined fortress of Louisburg, in Cape Breton Island,
the subsidence appears to have been at least four or five
feet since the middle of the eighteenth century. At
Lichtenfels (latitude 63°), in Greenland, according to
the Danish surveys, the sinking of the land since 1789
has amounted to six or eight feet. In the basin of
Hudson Bay, however, the observations of Dr. Bell
show that the re-elevation from the Champlain submergence is still in progress, its rate, according to his
estimate, reaching probably five to seven feet during
each century.

Turning to the glaciated regions of Europe, we find,
as in North America, that the countries which were ice-
covered had a much greater altitude before the ice accumulation, as shown by fiords. During the later part of
the Tertiary era and up to the Glacial period the larger
portion of the British Isles and all of Scandinavia
suffered vast denudation, with erosion of fiords and
channels that on the borders of Scotland and in the
Hebrides are now submerged 500 to 800 feet beneath
the sea. Much higher preglacial elevation is known for
Scandinavia, the principal centre of outflow of the European icefields. Jamieson notes the depth of the
Christiania Fiord as 1,380 feet, of the Hardanger Fiord
2,624 feet, and of Sogne Fiord, the longest in Norway,
4,080 feet.*

Afterward, when the European ice sheet was being
melted away, its area was mostly depressed somewhat
below its present level. The supposed great submergence, however, up to 1,200 and 1,500 feet, or more,

---

* Geol. Magazine, III, vol. viii, pp. 387–392, September, 1891.

which has been claimed by British geologists for northern Wales, northwestern England, and a part of Ireland, on the evidence of marine shells and fragments of shells in glacially transported deposits, is shown by Belt, Goodchild, Lewis, Kendall, and others, to be untenable. Indeed, these fossils, not lying in the place where they were living, give no proof of any depression of the land, since they have been brought by currents of the ice sheet moving across the bed of the Irish Sea. But it is clearly known by other evidence, as raised beaches and fossiliferous marine sediments, that large areas of Great Britain and Ireland were slightly depressed under their burden of ice, and have been since uplifted to a vertical extent, ranging probably up to a maximum of about 300 feet.

In Scandinavia the valuable observations and studies of Baron Gerard de Geer have supplied lines of equal depression of the land at the time of the melting away of the ice. This region of greatest thickness of the European ice sheet is found to have been depressed to an increasing extent from the outer portions toward the interior. The lowest limit of the submergence, at the southern extremity of Sweden, is no more than 70 feet above the present sea level, and in northeastern Denmark it diminishes to zero ; but northward it increases to an observed amount of about 800 feet on the western shore of the Gulf of Bothnia, near latitude 63°. Along the coast of Norway it ranges from 200 feet to nearly 600 feet, excepting far northward, near North Cape, where it decreases to about 100 feet. In proportion with this observed range of the subsidence on the coast of Scandinavia, its amount in the centre of the country was probably 1,000 feet.

A very interesting history of the postglacial oscilla-

tions of southern Sweden has been also ascertained by
Baron de Geer, which seems to be closely like the postglacial movements of the northeastern border of North
America.   As on our Atlantic coast, the uplift from the
Champlain submergence in that part of Sweden raised
the country higher than now.   The extent of this uplift appears to have been about 100 feet on the area
between Denmark and Sweden, closing the entrance to
the Baltic Sea, which became for some time a great
fresh-water lake.   After this another depression of that
region ensued, opening a deeper passage into the Baltic
than now, giving to this body of brackish water a considerably higher degree of saltness than at present, with
the admission of several marine molluscs, notably *Lito-
rina litorea* L., which are found fossil in the beds
formed during this second and smaller submergence, but
are not living in the Baltic to-day.   Thus far the movements of southern Sweden are paralleled by the postglacial oscillations of New England and eastern Canada,
but a second uplifting of this part of Sweden is now
taking place, whereas no corresponding movement has
begun on our Atlantic border.   It seems to be suggested, however, that it may yet ensue.   The subsidence
has ceased or become exceedingly slow in eastern New
England, while it still continues at a measurable rate in
New Jersey, Cape Breton Island, and southern Greenland.

So extensive agreement on opposite sides of the Atlantic in the oscillations of the land while it was ice-
covered, and since the departure of the ice sheets, has
probably resulted from similar causes—namely, the
pressure of the ice weight bearing down the land from
its great preglacial altitude, and the resilience of the
earth's crust from the subsidence when it was unbur-

dened. The restoration of isostatic equilibrium in each country is attended by minor oscillations, the conditions requisite for repose being overpassed by the early re-elevation of outer portions of each of these great glaciated areas.

In view of this harmony in the epeirogenic movements of the two continents during the Glacial and Recent periods, it seems evident that the close of the Ice Age was not long ago, geologically speaking, for equilibrium of the disturbed areas has not yet been restored. Furthermore, the close parallelism in the stages of progress toward repose indicates nearly the same time for the end of the Glacial period on both continents, and approximate synchronism in the pendulumlike series of postglacial oscillations.

The temperate Miocene climate of the arctic regions seems probably referable chiefly to a much freer oceanic circulation than now, permitting more of the warmth of the tropics to be carried by marine currents into the circumpolar area, which also may have had generally deeper seas and less land to be the gathering ground of ice sheets. In the ensuing Pliocene times the northern lands appear to have been gradually elevated and to have become continuous, before the Ice age, from North America across the area of the shallow Behring Sea and Strait to northern Asia, and from Norway across the now submarine Scandinavian plateau, as it has been called, having maximum depths of 1,500 to 1,700 feet, to the Färöe Islands, Iceland, and Greenland.

If we refer the erosion of the channels of Baffin Bay and the arctic archipelago west of Greenland to late Miocene and early Pliocene time, that extensive tract may well have subsided to its present height, or lower, producing the continuous water area of Baffin Bay, Smith

Sound, and Kennedy and Robeson Channels, before the late Pliocene and Pleistocene epochs of a nearly uniform circumpolar flora and fauna, which present, however, a considerable contrast, as was noted in foregoing chapters, on the opposite sides of Baffin Bay. Previous to the general Champlain subsidence of the northern ice-burdened lands, by which the reign of cold and ice in now temperate latitudes was brought rapidly to an end, and, indeed, previous to the Ice age itself, but after the water barrier of Baffin Bay existed, the hardier plants of Scandinavia and certain species of the land animals, birds, and insects of northern Europe had migrated across the now submerged plateau to Greenland, giving the remarkable European affinities of its flora and fauna.

Many species of plants and animals, probably including the reindeer and musk ox, survived even through the Glacial period on the low shore lines and the high nunataks of Greenland, which did not become wholly ice enveloped. In later times, since the icefields have been restricted on their borders, these hardy representatives of the richer preglacial fauna and flora have, in many instances, extended their range along the yet partially icebound coasts of Greenland, both around its northern and southern ends; and the Eskimos have come to share with the reindeer, seals, and polar bear the stern wintry climate of this mountain-girdled, icy land.

The native American peoples, now generally considered a division of the Mongolian race, appear to have migrated to our continent from northeastern Asia during the early Quaternary time of general uplift of northern regions which immediately preceded the Ice age. Then land undoubtedly extended across the present area of Behring Sea. The width of Behring Strait is only 28 miles, and its greatest depth is only 170 feet. Northward from this strait the Arctic Ocean is very shallow, and the 40-fathom or 240-foot submarine contour line is first reached at a distance of 400 to 500 miles. Similarly, on the south, the 40-fathom contour lies about 500 miles distant from the strait, passing close south of the Pribilof Islands; but southwestward from Behring Strait this line is reached at a distance of 250 miles. This large area of shallow sea was probably all land in the early Pleistocene or Lafayette epoch of general northern land elevation. During the Postglacial period, however, since the culmination of the Champlain subsidence, the region of Behring Strait, according to Dall, has undergone considerable re-elevation, so that the water there is now shallower than in the Champlain epoch. The many divergent branches of the American peoples, and their remarkable progress toward civilization in Mexico, Central America, and Peru, before the discovery by Columbus, indicate for this division of mankind probably almost as great antiquity as in the eastern hemisphere, where many lines of evidence point to the origin and dispersion of men at a time far longer ago than the 6,000 to 10,000 years which measure the Postglacial period.

CHAPTER XIII.

THE CAUSES OF THE ICE AGE.

WHAT were the causes of the accumulation of the ice sheets of the Glacial period? Upon their areas warm or at least temperate climates had prevailed during long foregoing geologic ages, and again at the present time they have mostly mild and temperate conditions. The Pleistocene continental glaciers of North America, Europe, and Patagonia have disappeared; and the later and principal part of their melting was very rapid, as is known by various features of the contemporaneous glacial and modified drift deposits, and by the beaches and deltas of temporary lakes that were formed by the barrier of the receding ice sheets. Can the conditions and causes be found which first amassed the thick and vastly extended sheets of land ice, and whose cessation suddenly permitted the ice to be quickly melted away?

Two classes of theories have been presented in answer to these questions. In one class, which we will first consider, are the explanations of the climate of the Ice Age through astronomic or cosmic causes, comprising all changes in the earth's astronomic relationship to the heat of space and of the sun. The second class embraces terrestrial or geologic causes, as changes of areas of land and sea, of oceanic currents, and altitudes of continents, while otherwise the earth's relations to external sources of heat are supposed to have been practically as now, or not to have entered as important factors in the problem.

It has been suggested that, as the sun and his planets are believed to be moving forward together through space, the Glacial period may mark a portion of the pathway of the solar system where less heat was supplied from the stars than along the earlier and later parts of this pathway. To this suggestion it is sufficient to reply, that the researches of Prof. S. P. Langley, now Secretary of the Smithsonian Institution, show that at the present time no appreciable measure of heat comes to us in that way, and that probably not so much as one degree of the average temperature of the earth's climates was ever, within geologic times, so received from all other sources besides the sun and the earth's own internal heat. Concerning the latter, also, it is well ascertained that during at least the Mesozoic, Tertiary, and Quaternary eras it has affected the climatic average by no more than a small fraction of a degree.

Others have suggested that the sun's heat has varied, and that the Ice Age was a time of diminished solar radiation. To this we must answer that during the centuries of written history, and especially during the past century of critical investigations in terrestrial and solar physics, no variations of this kind have been discovered. Such a cause of the glacial accumulations would have enveloped Alaska and Siberia with ice sheets and their drift deposits. The anomalous geographic distribution of the drift forbids this hypothesis.

Among all the theories of the causes of the Glacial period, the one which has attracted the most attention, not only of geologists but also of physicists and astronomers, was thought out by Dr. James Croll, and published in magazine articles during the years 1864 to

23

1874, and is most fully stated in his work entitled
Climate and Time (1875). Dr. Croll's theory, which
also has been very ably advocated by Prof. James Geikie
in The Great Ice Age (1874, 1877, and 1894), and by
Sir Robert S. Ball in The Cause of an Ice Age (1891),
attributes the accumulation of ice sheets to recurrent
astronomic cycles which bring the winters of each polar
hemisphere of the earth alternately into aphelion and
perihelion each 21,000 years during the periods of maximum eccentricity of the earth's orbit. Its last period
of this kind was from about 240,000 to 80,000 years ago,
allowing room for seven or eight such cycles and alternations of glacial and interglacial conditions. The
supposed evidence of interglacial epochs, therefore, gave
to this theory a wide credence; but the uniqueness of
the Glacial period in the long geologic record, and the
recent determinations of the geologic brevity of the time
since the ice sheets disappeared from North America
and Europe, make it clear, in the opinions even of some
geologists who believe in a duality or plurality of Quaternary glacial epochs, that not astronomic but geographic causes produced the Ice Age. From the
meteorologists' standpoint this astronomic explanation
of a formerly glacial climate in now temperate latitudes
has been alternately defended and denied, just as geologists have been divided in respect to its applicability to
the history of the Glacial period.

Many eminent glacialists, as James Geikie, Wahn-
schaffe, Penck, De Geer, Chamberlin, Salisbury, Shaler,
McGee, and others, believe that the Ice Age was com-
plex, having two, three, or more epochs of glaciation,
divided by long interglacial epochs of mild and temperate climate, when the ice sheets were entirely or
mainly melted away. Prof. Geikie claims six distinct

glacial epochs, as indicated by fossiliferous beds lying between deposits of till or unstratified glacial drift, and by other evidences of great climatic changes.   Mr. McGee, in the United States, recognises at least three glacial epochs.   On the other hand, the reference of all the glacial drift to a single epoch of glaciation, with moderate oscillations of· retreat and readvance of the ice border, is thought more probable by Dana, Hitchcock, and the present authors in America, Prestwich, Lamplugh, and Kendale in England, Falsan in France, Holst in Sweden, and Nikitin in Russia.   The arguments supporting this opinion are set forth by Prof. G. F. Wright in The Ice Age in North America (1889) and Man and the Glacial Period (1892), and especially in articles in the American Journal of Science for November, 1892, and March, 1894.

In accordance with Dr. Croll's astronomic theory, glacial periods would be expected to recur with geologic frequency, whenever the earth's orbit attained a stage of maximum eccentricity, during the very long Tertiary and Mesozoic eras, which together were probably a hundred times as long as the Quaternary era in which the Ice Age occurred.   But we have no evidence of any Tertiary or Mesozoic period of general glaciation in circumpolar and temperate regions, although high mountain groups or ranges are known to have had local glaciers.   Not until we go back to the Permian period, closing the Palæozoic era, are numerous and widely distributed proofs of very ancient glaciation encountered. Boulder-bearing deposits, sometimes closely resembling till and including striated stones, while the underlying rock also occasionally bears glacial grooves and striæ, are found in the Carboniferous or more frequently the Permian series in Britain, France, and Germany, Natal,

India, and southeastern Australia. In Natal the striated glacier floor is in latitude 30° south, and in India only 20° north of the equator. During all the earth's history previous to the Ice Age, which constitutes its latest completed chapter, no other such distinct evidences of general or interrupted and alternating glaciation have been found; and just then, in close relationship with extensive and repeated oscillations of the land, and with widely distant glacial deposits and striation, we find a most remarkable epoch of mountain-building, surpassing any other time between the close of the Archæan era and the Quaternary.

Alfred Russel Wallace therefore concludes, in his work on Island Life, that eccentricity of the earth's orbit, though tending to produce a glacial period, is insufficient without the concurrence of high uplifts of the areas glaciated. He thinks that the time of increased eccentricity 240,000 to 80,000 years ago was coincident with great altitude of northwestern Europe, North America, and Patagonia, which consequently became covered by ice sheets; but that such previous times of eccentricity, not being favoured by geographic conditions, were not attended by glaciation. The recentness of the Ice Age, however, seems to demonstrate that eccentricity was not its primary cause, and to bring doubt that it has exerted any determining influence in producing unusual severity of cold either during the Pleistocene or any former period.

In various localities we are able to measure the present rate of erosion of gorges below waterfalls; and the length of the postglacial gorge, divided by the rate of recession of the falls, gives approximately the time since the Ice Age. Such measurements of the gorge and Falls of St. Anthony by Prof. N. H. Winchell show the length

of the Postglacial or Recent period to have been about
8,000 years; and from the surveys of Niagara Falls, Mr.
G. K. Gilbert estimated it to have been 7,000 years, more
or less. From the rates of wave-cutting along the sides
of Lake Michigan and the consequent accumulation of
sand around the south end of the lake, Dr. E. Andrews
computed that the land there became uncovered from its
ice sheet not more than 7,500 years ago. Prof. G. F.
Wright obtains a similar result from the rate of filling
of kettle-holes among the gravel knolls and ridges called
kames and eskers, and likewise from the erosion of valleys by streams tributary to Lake Erie; and Prof. B. K.
Emerson, from the rate of deposition of modified drift
in the Connecticut Valley, at Northampton, Mass.,
thinks that the time since the Glacial period can not
exceed 10,000 years. An equally small estimate is also
indicated by the studies of Gilbert and Russell for the
time since the last great rise of the Quaternary Lakes
Bonneville and Lahontan, lying in Utah and Nevada,
within the arid Great Basin of interior drainage, which
are believed to have been contemporaneous with the
great extension of ice sheets upon the northern part of
the North American continent.

Prof. James Geikie maintains that the use of palæo-
lithic implements had ceased, and that early man in
Europe made neolithic (polished) implements, before
the recession of the ice sheet from Scotland, Denmark,
and the Scandinavian peninsula: and Prestwich suggests that the dawn of civilization in Egypt, China, and
India may have been coeval with the glaciation of northwestern Europe. In Wales and Yorkshire the amount
of denudation of limestone rocks on which drift boulders lie has been regarded by Mr. D. Mackintosh as
proof that a period of not more than 6,000 years has

elapsed since the boulders were left in their positions. The vertical extent of this denudation, averaging about six inches, is nearly the same with that observed in the southwest part of the Province of Quebec by Sir William Logan and Dr. Robert Bell, where veins of quartz, marked with glacial striæ, stand out to various heights not exceeding one foot above the weathered surface of the inclosing limestone.

From this wide range of concurrent but independent testimonies, we accept it as practically demonstrated that the ice sheets disappeared only 6,000 to 10,000 years ago. It is therefore manifestly impossible to ascribe their existence to astronomic causes which ceased 80,000 years ago, as is done by Croll's theory. Instead, we may believe, with Prestwich, that the Ice Age not only terminated, but began, after the end of the last period of maximum eccentricity of the earth's revolution around the sun.

Another astronomic theory, which assigns a date and duration of the Glacial period from about 24,000 to 6,000 years ago, agreeing nearly with Prestwich's estimate of its time, has been brought forward by Major-General A. W. Drayson, who first published it in the Quarterly Journal of the Geological Society of London for 1871, and latest in a volume entitled Untrodden Ground in Astronomy and Geology (1890). This theory asserts that the earth's axis during a cycle of about 31,000 years varies 12° in its inclination to the plane of the ecliptic or path of the earth around the sun. In this long cycle the axis and poles of the earth are thought to describe a circle in the heavens, with its centre 6° from the pole of the ecliptic. At present the obliquity of the ecliptic or angle between its plane and that of the earth's equator is about 23$\frac{1}{2}$°, which, therefore, is the distance of the arctic and antarctic circles from the poles; and this, according to Drayson's computations, is nearly their minimum distance. He claims that this obliquity of the ecliptic, which gives the distance of the arctic circles from the poles and of the tropics from the equator, about 5,000 years ago was some 2° more than now; that 7,500 years ago it was increased 6½° more than at present; that its maximum, nearly 12° more than at present, was about 13,500 B. C.; and that the beginning of this latest cycle of variation in the widths of the intertropical and polar zones was about 31,000 years ago. During the middle portion of the cycle, Drayson affirms that the arctic circle reached approximately to 54° north latitude, and that the resulting climatic changes caused the Ice Age.

It is true that the obliquity of the ecliptic varies slightly, and is at present decreasing about an eightieth part of a degree in 100 years. Sir John Herschel computed, however, that its limit of variation during the last 100,000 years has not exceeded 1° 21′ from its mean, although for a longer time in the past, as millions of years, it may range three or four degrees on each side of the mean. The portion of the present cycle of variation, which is used as the basis of this theory, seems insufficient to establish its conclusion of a wide range of obliquity; but, even if this were true, the same arguments forbid its application to account for the Glacial period as are urged by Gilbert, Chamberlin, and Le Conte in their dissent from Croll's theory.* These objections consist in the absence of evidence of glaciation during the long history of the earth previous to the Ice Age, excepting near the end of Palæozoic time, and the

---

* The Ice Age in North America. pp. 439, 440.

unsymmetric geographic areas of the ice sheets, northern
Asia and Alaska having not been ice-enveloped. According to Drayson, astronomic conditions capable of
producing an ice age have recurred every 31,000 years;
but geologists have recognised no other time of glaciation of large areas besides the Quaternary and Palæozoic
ice ages, which were divided probably by 10,000,000 or
15,000,000 years.

A third and much different theory, dependent on
the earth's astronomic relationship, is the suggestion
first made in 1866 by Sir John Evans, that, while the
earth's axis probably remained unchanged in its direction, a comparatively thin crust of the earth may have
gradually slipped as a whole upon the much larger nu-
cleal mass so that the locations of the poles upon the
crust have been changed, and that the Glacial period
may have been due to such a slipping or transfer by
which the regions that became ice-covered were brought
very near to the poles.*   The same or a very similar
view has been recently advocated by Dr. Fridtjof Nansen, who writes: " The easiest method of explaining a
glacial epoch, as well as the occurrence of warmer climates in one latitude or another, is to imagine a slight
change in the geographical position of the earth's axis.
If, for instance, we could move the north pole down to
some point near the west coast of Greenland, between
60° and 65° north latitude, we could, no doubt, produce
a glacial period both in Europe and America." †

Very small changes of latitude which had been de-
tected at astronomical observatories in England, Ger-

---

* Proceedings of the Royal Society of London, vol. xv, pp. 46–
54. February 28, 1866.

† The First Crossing of Greenland (1890), vol. ii, p. 454.

many, Russia, and the United States, seemed to give
some foundation for this theory, which in 1891 was regarded by a few American glacialists as worthy of attention and of special investigation by astronomers, with
temporary establishment of new observatories for this
purpose on a longitude about 180° from Greenwich or
from Washington. During the year 1892, however, the
brilliant discoveries by Dr. S. C. Chandler of the periods and amounts of the observed variations of latitude, showing them to be in two cycles (respectively of
twelve and fourteen months), with no appreciable secular change, forbid reliance on this condition as a cause,
or even as an element among the causes, of the Ice Age.
This theory is now entirely out of the field. Sir Robert
S. Ball, after reviewing Dr. Chandler's investigations,
estimates that the place of the pole since the Glacial
period, and from even earlier geologic times, has been
without greater changes of position than would lie inside the area of a block or square inclosed by the intersecting streets of a city.

The most recent mathematical investigation of the
effects of the unequal amounts of solar heat received by
different portions of the earth's surface, under varying
astronomic conditions, was published last year by Dr.
George F. Becker, of the United States Geological Survey, who sums up his results, differing antipodally from
the views of the late Dr. Croll, as follows : *

I began this inquiry without the remotest idea as to what conclusion would be reached. At the end of it I feel compelled to
assert that the combination of low eccentricity and high obliquity
will promote the accumulation of glacial ice in high latitudes
more than any other set of circumstances pertaining to the earth's
orbit. It seems to me that the glacial age may be due to these

_____

* Am. Jour. Sci., III. vol. xlviii, pp. 95–113. August, 1894.

conditions in combination with a favourable disposition of land
and water. This theory implies, or rather does not exclude, simul-
taneous glaciation in both hemispheres. It does not imply that
the ice age should last only ten or twelve thousand years. If the
conditions here suggested are correct, variations in the disposition
of land and water may have determined intervals of glaciation,
not necessarily the same ones in New England and the basin of
the Mississippi; and there may have been considerable time
differences in the inception or the cessation of glaciation in various regions. It is not needful to assume that the glaciation of
the Sierra Nevada either began or ended synchronously with the
ice age in New England. The date at which a minimum of eccentricity last coincided with a maximum of obliquity can almost
certainly be determined. According to Stockwell, the obliquity
has been diminishing for the past 8,000 years, and was within 21
minutes of its maximum value at the beginning of that time.
According to Leverrier, the eccentricity passed through a minimum 40,000 years ago, the value being then about two thirds of
the present one. So far as I know, the obliquity has not been
computed beyond 8,000. This can of course be done for Stockwell's value of the masses of the planets, or for newer and better
ones. All the indications seem to be that within thirty or forty
thousand years conditions have occurred, and have persisted for a
considerable number of thousand years, which would favour glaciation on the theory of this paper.

We come now to the wholly terrestrial or geologic
theory of the causes of the Ice Age, which, in terms
varying with increasing knowledge, has been successively
advocated by Lyell, Dana, Le Conte, and others, including the authors of the present volume. According to
this explanation, the accumulation of the ice sheets was
due to uplifts of the land as extensive high plateaus, receiving snowfall throughout the year. It may therefore
be very properly named the epeirogenic theory.

In the first edition of the Principles of Geology
(1830), Lyell pointed out the intimate dependence of
climate upon the distribution of areas of land and water

and upon the altitude of the land. In 1855, Dana, rea-
soning from the prevalence of fiords in all glaciated
regions, and showing that these are valleys eroded by
streams during a formerly greater elevation of the land
previous to glaciation, and from the marine beds of the
St. Lawrence Valley and basin of Lake Champlain belonging to the time immediately following the glaciation, announced that the formation of the drift in North
America was attended by three great continental movements: the first upward, during which the ice sheet
was accumulated on the land; the second downward,
when the ice sheet was melted away; and the third,
within recent time, a re-elevation, bringing the land to
its present height. But with the moderate depth of the
fiords and submarine valleys then known, the amount of
preglacial elevation which could be thus affirmed was
evidently too little to be an adequate cause for the cold
and snowy climate producing the ice sheet. The belief
that this uplift was 3,000 feet or more, giving sufficiently
cool climate, as Prof. T. G. Bonney has shown, to cause
the ice accumulation, has been reached only within the
past few years through the discovery by soundings of
the United States Coast Survey, that on both the Atlantic and Pacific coasts of the United States submarine
valleys, evidently eroded in late Tertiary and Quaternary
time, reach to profound depths—2,000 to 3,000 feet
below the present sea level.

The evidence of this great preglacial epeirogenic uplift, both for North America and Europe, has been
stated in the preceding chapter. Just the same high
preglacial elevation and Late Glacial or Champlain subsidence, with recent re-elevation, are known also for
Patagonia by its abundant and deep fiords and by its
marine beds overlying the glacial drift to heights of sev-

eral hundred feet above the sea, as described by Darwin
and Agassiz.   On these three continental areas the
widely separated, chief drift-bearing regions of the earth
are found to have experienced, in connection with their
glaciation, in each case three great epeirogenic movements of similar character and sequence—first, a comparatively long-continued uplift, which in its culmination appears to have given a high plateau climate with
abundant snowfall, forming an ice sheet, whose duration
extended until the land sank somewhat lower than now,
leading to amelioration of the climate and the departure
of the ice, followed by re-elevation to the present level.
The coincidence of these great earth movements with
glaciation naturally leads to the conviction that they
were the direct and sufficient cause of the ice sheets and
of their disappearance ; and this conclusion is confirmed
by the insufficiency and failure of the other theories
which have been advanced to account for the Glacial
period.

The end of the Tertiary era and the subsequent Lafayette, Glacial, and Recent periods, have been exceptionally characterized by many great oscillations of continental and insular land areas.   Where movements of
land elevation have taken place in high latitudes, either
north or south, which received abundant precipitation
of moisture, ice sheets were formed ; and the weight of
these ice sheets seems to have been a chief cause, and
often probably the only cause, of the subsidence of these
lands and the disappearance of their ice.

Between epochs of widely extended mountain-building by plication the diminution of the earth's mass produces epeirogenic distortion of the crust, by the elevation
of certain large areas and the depression of others, with
resulting inequalities of pressure upon different portions

of the interior; and these effects have been greatest immediately before relief has been given by the formation of folded mountain ranges. There have been two epochs pre-eminently distinguished by extensive mountain plication, one occurring at the close of the Palæozoic era, and another progressing through the Tertiary and culminating in the Quaternary era, introducing the Ice Age. With the culminations of both of these great epochs of mountain-building, so widely separated by the Mesozoic and Tertiary eras, glaciation has been remarkably associated, and, indeed, the ice accumulation appears to have been caused by the epeirogenic and orogenic uplifts of continental plateaus and mountain ranges. * The earth's surface is probably now made much more varied, beautiful, and grand by the existence of many lofty mountain ranges than has been its average condition during the past long eras; but the magnificent Pleistocene icefields and glaciers have vanished, or are much diminished, excepting only in Greenland and on the antarctic continent.

---

* For a more extended discussion of the relationship of the earth's cooling and contraction, with movements of continental and mountain uplifts, and with the accumulation of ice sheets, see an appendix on Probable Causes of Glaciation, in Wright's Ice Age in North America, pp. 573–595.

CHAPTER XIV.

FROM the consideration of the movements of uplift
and subsidence of the lands on both sides of the North
Atlantic Ocean, which are believed to have caused the
envelopment of vast areas of these continents by ice
sheets like those now covering Greenland and the antarctic continent, we can advance to a completion of
our review of the Glacial period by noticing the several
stages of this period and the sequence of its history as
revealed by its marginal moraines and other drift deposits. Exploration of the European glacial drift by
two Americans, Prof. H. Carvill Lewis in the British
Isles and Prof. R. D. Salisbury in Germany, less than
ten years ago, laid the foundations for determining the
geologic equivalency of the successive parts of the drift
series in North America and Europe. Salisbury especially noted that the marginal moraines of northern
Germany lie, as in the United States, at some distance
back from the limits of the drift.

Studies by many observers have shown that on both
continents the border of the drift along the greater part
of its extent was laid down as a gradually attenuated
sheet; that the ice retreated and the drift endured much
subaërial erosion and denudation; that renewed accumulation and growth of the ice sheet, but mostly with-

348

out extending to its earlier limits, were followed by a general depression of these burdened lands, after which the ice again retreated, apparently at a much faster rate than before, with great supplies of loess from the waters of its melting; that moderate re-elevation ensued; and that during the farther retreat of the ice sheet prominent moraines were amassed in many irregular but roughly parallel belts, where the front at successive times paused or readvanced under secular variations in the prevailingly temperate and even warm climate, by which, between the times of formation of the moraines, the ice was rapidly melted away.

Such likeness in the sequence of glacial conditions probably implies contemporaneous stages in the glaciation of the two continents; and the present authors believe that it is rather to be interpreted as a series of phases in the work of a single ice sheet on each area than as records of several separated and independent epochs of glaciation, differing widely from one another in their methods of depositing drift. The latter view, however, is held by James Geikie, Penck, De Geer, and others in Europe; and it has been regarded as the more probable also for America by Chamberlin, Salisbury, McGee, and others.

Under this view Geikie distinguishes no less than eleven stages or epochs, glacial and interglacial, which he has very recently named,* since the publication last year of the new edition of his Great Ice Age, in which, however, they were fully described. These divisions of the Glacial period are as follows: 1, The Scanian or first glacial epoch; 2, The Norfolkian or first interglacial epoch; 3, The Saxonian or second glacial epoch;

---

* Journal of Geology, vol. iii, pp. 241-269, April–May, 1895.

4, The Helvetian or second interglacial epoch; 5, The Polandian or third glacial epoch; 6, The Neudeckian or third interglacial epoch; 7, The Mecklenburgian or fourth glacial epoch; 8, The Lower Forestian or fourth interglacial epoch; 9, The Lower Turbarian or fifth glacial epoch; 10, The Upper Forestian or fifth interglacial epoch; and, 11, The Upper Turbarian or sixth glacial epoch.

The earliest application of such geographic names to the successive stages and formations of the Ice Age appears to be that of Chamberlin in his two chapters contributed to the new third edition of Geikie's admirable work before mentioned, in which he names the Kansan, East Iowan, and East Wisconsin formations. For the second and third he has since adopted the shorter names, Iowan and Wisconsin. This classification he has also more recently extended, the interglacial stage and deposits between the Kansan and Iowan glacial drift or till formations being named Aftonian, and the Toronto interglacial formation, previously named, being referred, with some doubt, to an interval between the Iowan and Wisconsin stages. Chamberlin correlates, with a good degree of confidence, his Kansan stage of maximum North American glaciation with the maximum in Europe, which is Geikie's Saxonian epoch; the Aftonian stage as Geikie's Helvetian; the Iowan as the European Polandian; and the Wisconsin or moraine-forming stage of the United States as the Mecklenburgian, which was the stage of the "great Baltic Glacier" and its similarly well-developed moraines.*

According to the law of priority, the names of the Kansan, Iowan, and Wisconsin formations and stages

---

* Journal of Geology, vol. iii, pp. 270–277, April–May, 1895.

should also be applied to these European divisions of the Glacial series, for the studies of Geikie and Chamberlin show them to be in all probability correlative and contemporaneous. Figures 57 and 58 therefore employ these names for both our own continent and Europe, giving the boundaries of these formations as mapped in the Great Ice Age, and adding for the northeastern United States and Canada the Warren, Toronto, Iroquois, and St. Lawrence stages in the glacial recession, nearly as indicated in a recent article on the glacial representatives of the Laurentian lakes and on the Late Glacial or Champlain subsidence and re-elevation of the St. Lawrence river basin.*

Differing much from the opinions of Geikie, and less widely from those of Chamberlin, concerning the importance, magnitude, and duration of the interglacial stages, but agreeing with Dana, Hitchcock, Kendall, Falsan, Holst, Nikitin, and others, in regarding the Ice Age as continuous, with fluctuations but not complete departure of the ice sheets, our view of the history of the Glacial period, comprising the Glacial epoch of ice accumulation and the Champlain epoch of ice departure, may be concisely presented in the following somewhat tabular form. The order is that of the advancing sequence in time, opposite to the downward stratigraphic order of the glacial, fluvial, lacustrine, and marine deposits.

---

* American Journal of Science, III. vol. xlix. pp. 1–18, with map, by Warren Upham, January, 1895. Mr. Upham's first essay toward the discrimination of the two chief epochs and the successive stages of the Ice Age, as recognisable alike in North America and Europe, was given in the American Naturalist, vol. xxix. pp. 235–241, March, 1895.

24

EPOCHS AND STAGES OF THE GLACIAL PERIOD.

I. *The Glacial Epoch.*

1. THE CULMINATION OF THE LAFAYETTE EPEIROGENIC UPLIFT, affecting both North America and Europe, raised the glaciated areas to so high altitudes that they received snow throughout the year, and became deeply ice-enveloped. Valleys and fiords show that this elevation was 1,000 to 4,000 feet above the present height.

Rudely chipped stone implements and human bones in the plateau gravel of southern England, 90 feet and higher above the Thames, and the similar traces of man in high terraces of the Somme valley, attest man's existence there before the maximum stages of the uplift and of the Ice Age. America also had been already peopled, doubtless by preglacial migration from Asia across a land area in the place of the present shallow Behring Sea.

The accumulation of the ice sheets, due to snowfall upon their entire areas, was attended by fluctuations of their gradually extending boundaries, giving the Scanian and Norfolkian stages in Europe, the Albertan formation of very early glacial drift and accompanying gravels, recently described by Dr. George M. Dawson, in Alberta and the Saskatchewan district of western Canada, and an early glacial advance, recession, and readvance, in the region of the Moose and Albany Rivers, southwest of Hudson Bay. In that region, and westward on the Canadian plains to the Rocky Mountains, there seem to be thus three stages recognisable in the glacial results of the epeirogenic uplift, namely, the ALBERTAN STAGE of early ice accumulation, the SASKATCHEWAN STAGE of abundant melting and considerable retreat, and the

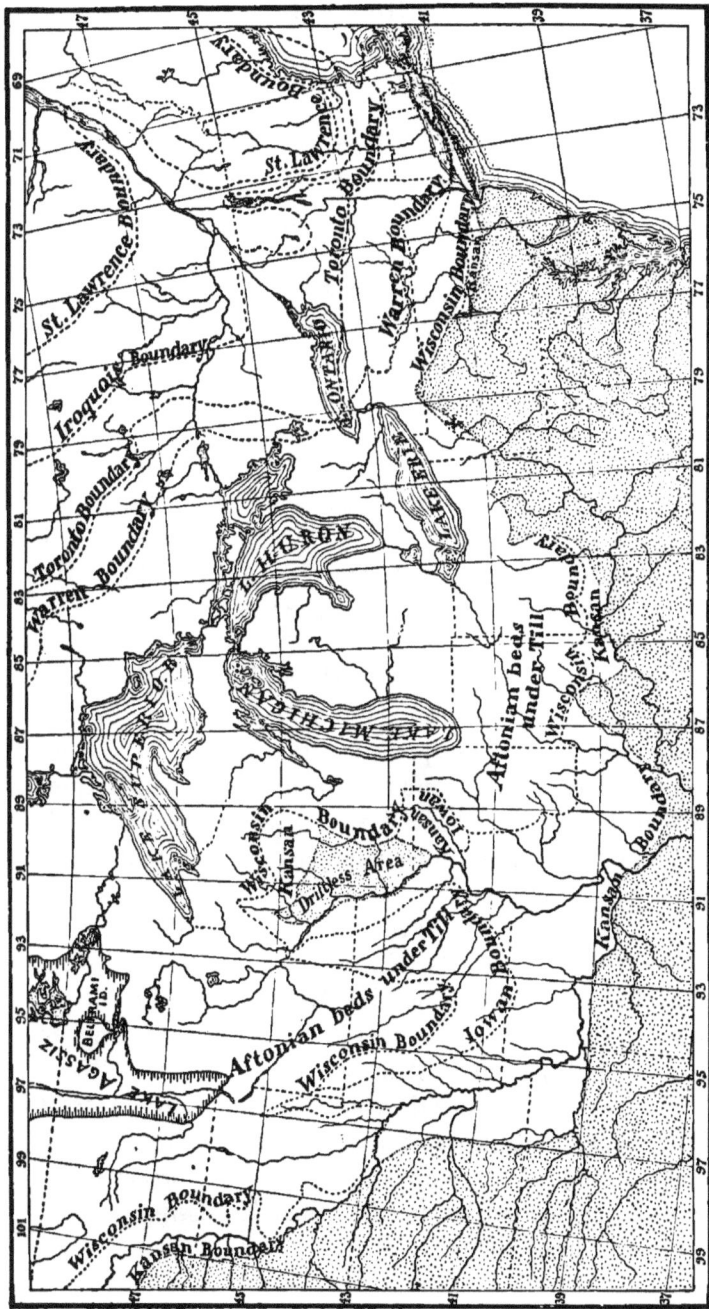

FIG. 57.—Stages in the recession of the North American ice sheet. Glaciated portion unshaded.

ensuing great Kansan growth of the continental icefields.

2. KANSAN STAGE.—Farthest extent of the ice
sheet in the Missouri and Mississippi river basins, and
in northern New Jersey. The Saxonian stage of maximum glaciation in Europe.

Area of the North American ice sheet, with its development on the arctic archipelago, about 4,000,000
square miles; of the Greenland ice sheet, then somewhat more extended than now, 700,000 square miles or
more, probably connected over Grinnell Land and Ellesmere Land with the continental ice sheet, the area of
Greenland being approximately 680,000 square miles,
and of its present ice sheet 575,000 square miles; of the
European ice sheet, with its tracts now occupied by the
White, Baltic, North, and Irish Seas, about 2,000,000
square miles.

Thickness of the ice in northern New England and
in central British Columbia, about one mile; on the
Laurentide highlands, probably two miles; in Greenland, as now, probably one mile or more, with its surface
8,000 to 10,000 feet above the sea; in portions of Scotland and Sweden, and over the basin of the Baltic Sea,
half a mile to a mile.

3. HELVETIAN OR AFTONIAN STAGE.—Recession of
the ice sheet from its Kansan boundary northward
about 500 miles to Barnesville, Minn., in the Red River
Valley; 250 miles or more in Illinois, according to
Leverett, but probably little between the Scioto River,
in Ohio, and the Atlantic coast, the maximum retreat
of that portion being 25 miles or more in New Jersey.
A cool temperate climate and coniferous forests up to
the receding ice border in the upper Mississippi region.
Much erosion of the early drift.

The greater part of the drift area in Russia permanently relinquished by the much-diminished ice sheet, which also retreated considerably on all its sides.

During this stage the two continents probably retained mainly a large part of their preglacial altitude. The glacial recession may have been caused by the astronomic cycle which brought our winters of the northern hemisphere in perihelion between 25,000 and 15,000 years ago.*

4. IOWAN STAGE.—Renewed ice accumulation, covering the Aftonian forest beds, and extending again into Iowa, to a distance of 350 miles or more from its most northern indentation by the Aftonian retreat, and re-advancing about 150 miles in Illinois, while its boundary eastward from Ohio probably remained with little change.

The Polandian stage of renewed growth of the European ice sheet, probably advancing its boundaries in some portions hundreds of miles from the Helvetian retreat.

II. *The Champlain Epoch.*

5 CHAMPLAIN SUBSIDENCE; NEUDECKIAN STAGE. —Depression of the ice-burdened areas mostly somewhat below their present heights, as shown by fossiliferous marine beds overlying the glacial drift up to 300 feet above the sea in Maine, 560 feet at Montreal, 300 to 400 feet from south to north in the basin of Lake Champlain, 300 to 500 feet southwest of Hudson and James Bays, and similar or less altitudes on the coasts of British Columbia, the British Isles, Germany, Scandinavia, and Spitzbergen.

---

* American Geologist. vol. xv, pp. 201, 255, and 293, March, April, and May, 1895.

Glacial recession from the Iowan boundaries was rapid under the temperate (and in summers warm or hot) climate, belonging to the more southern parts of

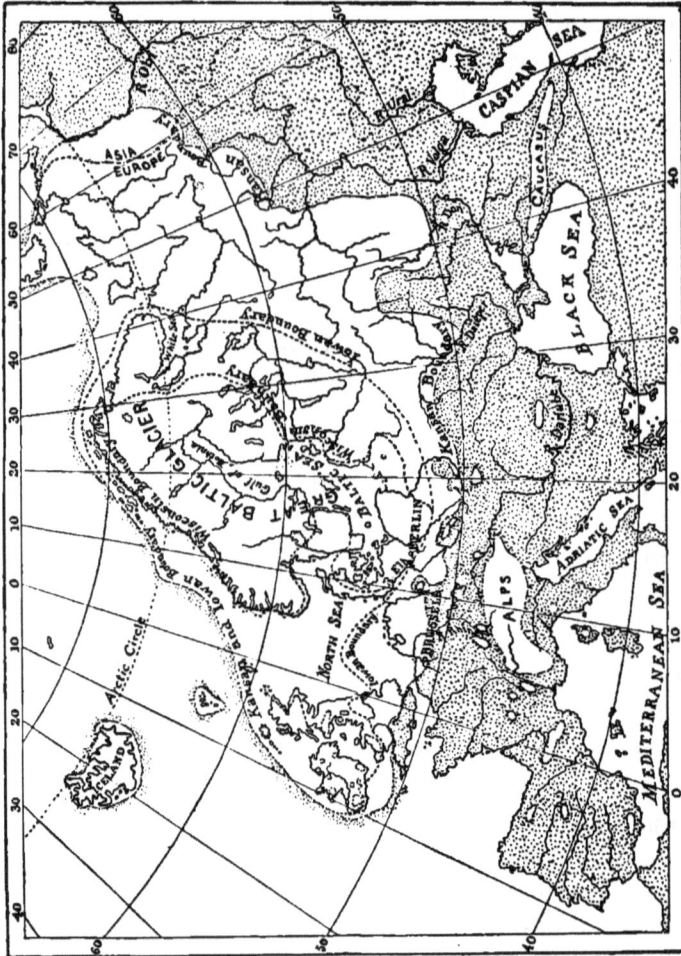

Fig. 58.—Stages in the recession of the European ice sheet. Glaciated portion unshaded.

the drift-bearing areas when reduced from their great preglacial elevation to their present height or lower. The finer portion of the englacial drift, swept down from the icefields by the abundant waters of their melting and of rains, was spread on the lower lands and along valleys in front of the departing ice, as the loess of the Missouri, the Mississippi, and the Rhine. Marine beds reaching to a maximum height of about 375 feet at Neudeck, in western Prussia, give the name of this stage.

6. WISCONSIN STAGE.—Moderate re-elevation of the land, in the northern United States and Canada advancing as a permanent wave from south to north and northeast; continued retreat of the ice along most of its extent, but its maximum advance in southern New England, with fluctuations and the formation of prominent marginal moraines; great glacial lakes on the northern borders of the United States.

The Mecklenburgian stage in Europe. Conspicuous moraine accumulations in Sweden, Denmark, Germany, and Finland, on the southern and eastern margins of the great Baltic glacier. No extensive glacial readvance between the Iowan and Wisconsin stages, either in North America or Europe.

7. WARREN STAGE.—Maximum extent of the glacial Lake Warren, held on its northeast side by the retreating ice border; one expanse of water, as mapped by Spencer, Lawson, Taylor, Gilbert, and others, from Lake Superior over Lakes Michigan, Huron, and Erie to the southwestern part of Lake Ontario; its latest southern beach traced east by Gilbert to Crittenden, New York, correlated by Leverett with the Lockport moraine.[*]

This and later American stages, all of minor impor-

---

* American Journal of Science, III, vol. 1, pp. 1–20, with map, July, 1895.

tance and duration in comparison with the preceding, can not probably be shown to be equivalent with Geikie's European divisions belonging in the same time. Successive boundaries of the receding American ice sheet are noted, as in Figure 57, in accordance with studies of the Laurentian series of glacial lakes.

8. TORONTO STAGE.—Slight glacial oscillations, with temperate climate nearly as now, at Toronto and Scarborough, Ontario, indicated by interbedded deposits of glacial drift and fossiliferous stratified gravel, sand, and clay.* Although the waning ice sheet still occupied a vast area on the northeast, and twice readvanced, with deposition of much boulder clay or till, during the formation of this fossiliferous drift series, the climate then, determined by the Champlain low altitude of the land, by the proximity of the large glacial Lake Algonquin, succeeding the larger Lake Warren, and by the eastward and northeastward surface atmospheric currents and courses of all storms, was not less mild than now. The trees whose wood is found in the interglacial Toronto beds now have their most northern limits in the same region.

9. IROQUOIS STAGE.—Full expansion of the glacial Lake Iroquois in the basin of the present Lake Ontario and northward, then outflowing at Rome, New York, to the Mohawk and Hudson Rivers. Gradual re-elevation of the Rome outlet from the Champlain subsidence had lifted the surface of Lake Iroquois in its western part from near the present lake level at Toronto to a height there of about 200 feet, finally holding this height during many years, with the formation of the well-developed Iroquois beach.

* American Geologist, vol. xv, pp. 285-291, May, 1895.

Between the times of Lakes Warren and Iroquois, the glacial Lake Lundy, named by Spencer from its beach ridge of Lundy's Lane,* probably had an outlet east to the Hudson by overflow across the slope of the highlands south of the Mohawk; but its relationship to the glacial Lake Newberry, named by Fairchild as outflowing to the Susquehanna by the pass south of Seneca Lake,† needs to be more definitely ascertained.

10. St. Lawrence Stage.—The final stage in the departure of the ice sheet, which we are able to determine from the history of the Laurentian lakes and St. Lawrence Valley, is approximately delineated in Figure 57, when the glacial Lake St. Lawrence, outflowing through the Champlain basin to the Hudson, stretched from a strait originally 150 feet deep over the Thousand Islands, at the mouth of Lake Ontario, and from the vicinity of Pembroke, on the Ottawa River, easterly to Quebec or beyond. As soon as the ice barrier which had held these glacial lakes was melted through, the sea entered the depressed St. Lawrence, Champlain, and Ottawa valleys; and subsequent epeirogenic uplifting has raised them to their present slight altitude above the sea level.

Later stages of the glacial recession are doubtless recognisable by moraines and other evidences, the North American ice sheet becoming at last, as it probably also had been in its beginnings, divided into three parts, one upon Labrador, another northwest of Hudson Bay, as shown by Tyrrell's observations, and a third upon the northern part of British Columbia. From com-

---

* American Journal of Science, III. vol. xlvii. pp. 207–211, with map, March, 1894.

† Bulletin, Geological Society of America, vol. vi, pp. 353–374, with map and plates, April, 1895.

parison with the glacial Lake Agassiz in the basin of the
Red River of the North and of Lake Winnipeg, whose
duration was probably only about a thousand years, the
whole Champlain epoch of land depression, the departure of the ice sheet because of the warm climate so restored, and most of the re-elevation of the unburdened
lands, appear to have required only a few (perhaps four
or five) thousand years, ending about five thousand years
ago. These late divisions of the Glacial period were far
shorter than its Kansan, Aftonian, and Iowan stages;
and the ratio of the Glacial and Champlain epochs may
have been approximately as ten to one. The term
Champlain conveniently designates the short closing
part of the Ice Age when the land depression caused
rapid though wavering retreat of the ice border, with
more vigorous glacial currents on account of the marginal melting and increased steepness of the ice front,
favouring the accumulation of many retreatal moraines
of very knolly and bouldery drift.

The Glacial period or Ice Age is thus found divisible
into two parts or epochs, the first or Glacial epoch being
marked by high elevation of the drift-bearing areas and
their envelopment by vast ice sheets, and the second or
Champlain epoch being distinguished by the subsidence
of these areas and the departure of the ice, with abundant deposition of both glacial and modified drift.
Epeirogenic movements, first of great uplift and later
of depression, are thus regarded as the basis of the two
chief time divisions of this period. Each of these epochs
is further divided in stages, marked in the Glacial epoch
by fluctuations of the predominant ice accumulation,
and in the Champlain epoch by successively diminishing
limits of the waning ice sheet, which, however, sometimes temporarily readvanced, inclosing stratified and

fossiliferous beds between the unstratified glacial deposits. The Ice Age still lingers in Greenland and in the Alaskan region of Mount St. Elias, from which, and from the Alps, our interpretations of the meaning of the North American and European drift formations are derived.

CHAPTER XV.

THE preceding survey of facts impressively shows that Greenland is not only an important object lesson in glacial geology, but an intricate puzzle as well. Contrary to all ordinary expectations, Greenland is maintaining for itself an independent glacial period long after glacial conditions have ceased to exist in the corresponding latitudes of other portions of the northern hemisphere. Cape Farewell is on the sixtieth parallel of north latitude, and from that point northward, for a distance of fifteen hundred miles, the conditions of the Glacial period continue without cessation. But with the exception of the glaciers coming down from the lofty heights of the St. Elias Alps, and of a limited number on the west side of Baffin Bay north of Hudson Strait, there are no glaciers north of this latitude in British America and Alaska, while in Europe the capitals of Norway and Russia lie almost exactly upon the sixtieth parallel, and only a limited number of glaciers is found to the northward in the Scandinavian Peninsula, they being absent largely even from Lapland, which stretches beyond the seventieth parallel. The extensive areas of Finland and of Russia, in Europe, north of the latitude of Cape Farewell, which were once deeply enveloped in glacial ice, are now fruitful fields supporting a large and enterprising population. To understand the reason of

362

these present diverse conditions is probably to unravel the main cause of the Glacial period.

It is interesting to note that the recent inventions for distributing heat throughout our houses by a skillfully adjusted system of warm-water pipes are but imitations of a plan followed by Nature from time imme-

FIG. 59.--Iceberg off the coast of Labrador.

morial. A study of the present climatic conditions of the globe shows that the agencies for the distribution of the sun's heat are as important in the production of results as is the direct action of the sun itself.* The Gulf Stream in the Atlantic Ocean trends eastward, by reason

* For references and a fuller statement of facts, see Wright's Ice Age in North America, pp. 419–432.

of the diurnal motion of the earth from west to east,
and, finding free access to the seas stretching northward
from Europe, transfers hither the heat of the tropics to
modify the present climate of that once-frozen region.
But if warm water from the south is permitted to enter
the Arctic Ocean, cold water from the north must move
southward from that area to take its place. By the
same diurnal motion of the earth which throws north-
erly flowing oceanic currents eastward, the southerly
flowing arctic currents are thrown westward, thus surrounding Greenland with their chilling influences.

It is significant also that the present narrowness and
shallowness of Behring Strait prevent the warm currents
of the Pacific from entering the Arctic Ocean from the
west, so that they can not now assist in ameliorating the
climate of the lands about the pole. Under these conditions it is easy to see that a moderate depression of
Behring Strait might admit sufficient heat from the
Pacific Ocean to change the whole aspect of the climates within the arctic circle, while a somewhat greater
elevation of the sea bottom along the line of Iceland
and the Färöe Islands between Greenland and Scotland
might shut off so much warmth from the arctic area as
greatly to extend the conditions favourable to the production of glaciers.

Theoretical considerations of this sort combine with
much positive evidence to give great cogency to the
theory that the Glacial period was caused mainly by
changes of the relative sea level in the northern part of the
northern hemisphere, since these changes of level would,
so to speak, turn the currents of warm water on and off
from the different areas, thus controlling their local temperature, as the householder does that of the different portions of the building which is warmed by modern methods.

FIG. 60.—Icebergs off Labrador, seen from the shore at a distance.

The comparative shallowness of all these northern seas favours this theory, since only moderate changes of level would be required to produce the results. The positive evidences of such changes of level have already been detailed with sufficient fullness, but may here profitably be summarized and somewhat enlarged.

Previous to the Glacial period warm climates did actually exist everywhere north of the arctic circle. At that time the vegetation of central Europe and of the Middle Atlantic States of America flourished in northern Greenland and in Spitzbergen, while there are abundant independent evidences that this was at a period when the arctic lands were greatly depressed. At the same time it is equally clear, from the fiords and other submerged channels, that just preceding the Glacial period there was an extensive elevation of all northern land. This coincidence of late Tertiary land elevation and the great increase of glaciers is too extensive to have been accidental, and, both in its effect upon the distribution of oceanic currents and upon temperature by virtue of elevation, it is just the combination which should be expected to produce a Glacial period.

There is considerable difference of opinion as to what would be the character of Greenland if the ice were melted from it. Some maintain that the mountains upon the east and west coasts border a continent of little or moderate elevation in the interior, being in this respect analogous to most other large areas of land. Indeed, some suggest that much of the interior may be below the sea level, and that the real Greenland is but an archipelago. In this case the great height of the interior is produced by the accumulation of snow, such as we suppose at one time obliterated the North Sea, and joined the mountains of Scandinavia with those of the

British Isles in one continuous glacial sheet. The great size of the icebergs floating off from the Greenland coast gives much countenance to this view.

The persistence of the ice sheet over Greenland is also, perhaps, an effect of such a condition of things, since the great mass of ice implied in this theory would act as a powerful conservative agency to resist moderate changes in climate, while the great thickness of the accumulation would add to the arctic character of the climate. The small precipitation in Greenland—commonly stated to be only about ten inches annually on and near the coast—renders it quite probable that if the ice were once melted away, it would not, under present conditions, accumulate again. On the high surface of the ice sheet, however, according to the experience of Nansen and Peary, the snowfall is doubtless far more than along the coast.

On the other hand, it is maintained that there is not sufficient positive evidence of so thick an accumulation of ice as this theory implies. The great size of icebergs proves only a corresponding thickness of ice in the few depressions along the seashore through which the great glaciers discharge their contents. It is maintained that this may not be inconsistent with a much thinner accumulation over the most of the interior. Upon this theory Greenland may differ from the Scandinavian peninsula chiefly in the large extent of its *névé* fields and in their slightly higher altitude; so that the elevation of the land itself, surrounded as it is by water, may contribute largely to the production of glacial conditions.

The distribution and preservation of plants and animals in Greenland also strongly confirms the theory of a former extensive elevation of these northern regions.

25

FIG. 61.—View of the inland ice, east of the outskirts, near Sukkertoppen.

Asa Gray and others have shown that the affinity of the plants of southern Greenland with those of Europe is such as to make it probable that they emigrated directly from Europe, rather than by the longer route across Asia and North America.* Davis Strait seems to have been a more effectual barrier to the emigration of plants than was the North Atlantic on the east of Greenland. This would imply that the elevation of the bed of the North Atlantic is more certainly proved, or that it was longer continued than that of Davis Strait or Baffin Bay; or possibly it may prove simply that, from being freer of ice, it was more available for the passage of plants and animals.

But the question arises, How could the plants and animals which had arrived on the coast of Greenland survive the rigours of the Glacial period on the borders of a region which is covered with glacial ice at the present day? The answer is found partly in the conclusions to which recent investigations have led, that, much as the regions of the North Atlantic were elevated, the glaciers of Greenland did not extend so as to cover all the border which was raised above sea level. Indeed, while it seems evident that nearly all of the present border of Greenland was covered by glacial ice, the mountain peaks were never wholly enveloped, but projected as *nunataks* above the icy wastes. On these, as Mr. Upham surmises, many plants may successfully have maintained their struggle for existence.

But more probably we may picture Greenland of the Glacial period to ourselves as surrounded by wide, outlying, unglaciated areas of lowland but little above sea level, extending around the border, especially to the eastward.

* See above, pp. 201, 202.

Here the reindeer and musk ox, and some other animals, with all the existing plants of Greenland, may easily have maintained their existence until the ameliorating conditions of the climate caused the ice to withdraw toward its present limits, at the same time that the subsidence of the land curtailed their extension outward. In no other way does it seem possible to imagine how the reindeer and musk-ox (even if they could cross Smith Sound from America) could pass the icy barriers of Melville Bay and of the Humboldt glacier to reach the pasturing grounds of northern, eastern, and southern Greenland.

The evidence that the mountains on the border of Greenland were never wholly enveloped by ice consists largely in the contrast between their appearance and that of the mountains on the coast of Labrador. As already remarked,* there are in Labrador no sharp peaks, but everywhere the mountains present a smooth and flowing outline against the sky; while on the Greenland coast the prevailing feature of the landscape is that of sharp, needlelike peaks, such as would do credit to the high Alps, or to the central portion of the Rocky Mountain range.

The reason for this contrast can not be found in any dissimilarity between the character of the rocks in the two regions; for, as already noted, they are of the same age and essentially alike, being chiefly granitic and gneissoid in character, belonging to the Archæan period. The most satisfactory explanation, therefore, seems to be that the Labrador coast has been more completely enveloped in glacial ice than the Greenland coast has ever been, resulting in its being planed down to a more uni-

* See page 29.

form level; for it is evident that the erosive action of an ice sheet is prevailingly horizontal, while that of water is, in its earlier stages at least, when the land is much elevated, largely vertical, concentrating its force along the lowest lines, producing gorges and cañons, which are gradually enlarged into valleys separated by mountain peaks.

Mr. Walcott, however, the present director of the United States Geological Survey, believes[*] that the shaping of present land surfaces, even in these northern regions, is largely due to the action of ordinary erosive agencies in pre-Cambrian times, when the entire Archæan area was brought down approximately to a base-level. This is inferred in part from the situation of the Cambrian strata which appear in various places in the valley of the St. Lawrence and in Newfoundland, and, we may add, in Labrador as well.[†] According to Mr. Walcott's observations, whenever the Cambrian or Silurian rocks in the regions mentioned rest on the older crystallines, the surface beneath them is essentially of the same type as the general surface of the areas of crystalline rocks beyond them.

But the evidence of a long-continued elevation and subsequent glaciation of the area west of Davis Strait is beyond question. This has already been largely presented in our chapter upon Pleistocene Changes of Level, but the full strength of the case calls for some additional facts and discussion. Besides the great depths of the Saguenay cañon and of the submerged channel extending from it through the Gulf of St. Lawrence to

---

[*] See Twelfth Annual Report of the United States Geological Survey, pp. 546 *et seq.*

[†] See Chapter II.

the edge of the deep depression of the Atlantic basin,
which have already been adduced, we should add the
facts elsewhere noted concerning the depth of Hamilton
Inlet in Labrador, and of the lakes in the upper part of
Hamilton River below the Great Falls. Analogous facts
occur also along the eastern margin of Newfoundland,
which properly belongs to the same area. Grand Pond,
on this island, whose surface is 116 feet above tide, has a
depth of more than 1,000 feet, making its bottom 988
feet below sea-level. In Conception Bay, also, which is
almost like a fiord, the depth of the water is 840 feet.

To the evidence already adduced concerning the general glaciation of Labrador, we may add the similar
facts concerning Newfoundland, where glacial striæ are
found all along the eastern coast upon the highest of
the headlands. In the vicinity of St. John's the striæ
are very clearly marked in an east-and-west direction
upon the highest summits, 600 feet above the sea.
There seems, therefore, little reason to doubt the conclusion of Mr. Murray,[*] that the glacial phenomena of
Newfoundland belong to a general movement which filled
the Gulf of St. Lawrence and extended some distance out
upon the Atlantic plateau to the east of that island. Indeed, there is much reason to believe that the banks of
Newfoundland, so celebrated for their fisheries, are morainic deposits made during the maximum extension of
the ice which flowed off from the Labrador peninsula,[†] at a
time when the elevation was such as to raise this portion of the plateau above the water level.

_____

[*] Proceedings and Transactions of the Royal Society of Canada for 1882, sect. iv, pp. 55–76.

[†] See an article of G. F. Wright, in American Journal of Science, February, 1895, pp. 86–94.

Upon turning our attention to the Greenland coast, we have, as already shown, even more striking evidences of extensive elevations of the land in preglacial times, while it is equally clear that the accumulation of ice did not proceed far enough to envelop completely the highest mountain peaks which border the western coast. But that the ice sheet formerly extended much farther than it does at present, is evident from even a cursory examination. Referring only to my own observations, which, so far as I know, are in accordance with all others in southern Greenland, I noted at Ikamiut glacial striæ on the sides of the fiord several hundred feet above the water level pointing toward the open sea. Upon the margin of Isortok Fiord, also, glacial grooving and striation appeared on a magnificent scale near its mouth, which is now fifty miles from the head of the fiord. The grooves here were passing diagonally upward and across a projection of land separating the narrow portions of the fiord from the open sea and rising to a height of several hundred feet. Accumulations more nearly approaching a moraine than anything else I saw in Greenland occurred at this point. At Sukkertoppen, while there was no glacial till, but the rocks were perfectly bare and free from soil, there were at the same time, scattered over the low elevations, many boulders of a light colored granite, which were not local, but must have been brought from the interior. Beyond question, therefore, all the so-called "outskirts" of Greenland were formerly enveloped in ice, with the exception, possibly, of the higher mountain peaks which projected above it, as the nunataks of Dalager and Jensen do at the present time.*

---

* See above, pp. 264–267.

The complication of the phenomena, however, and their difficulty of interpretation, strikingly appear in the facts already noted by Prof. Chamberlin, that in the far north, near the seventy-sixth parallel of latitude, there are in close proximity the same contrasts between glaciated and unglaciated areas which we have noted respecting Labrador and the general coast of Greenland. Dalrymple Island, in latitude 76° 30', and near the Greenland coast, presents the needlelike contour of its mountain peaks characteristic of an unglaciated region; while fifty miles northwestward, the Carey Islands, consisting of almost the same identical rock (hornblendic gneiss), present the familiar graceful outlines character-istic of glaciated regions. The ice movement over the Carey Islands, however, was evidently from the north— that is, down Smith Sound—and not from the west coast of northern Greenland. This is shown both by the fact that the stoss sides are on the north, and that the erratics upon the islands, consisting of limestone, sandstone, shale, and quartzite, must have come from that direction.

At the same time, it is a suggestive commentary upon the local variations which may take place in connection with extensive ice movements, that within fifty miles north of the Carey Islands the sea is now more than thirteen hundred feet deep, and a small portion of the coast east of Dalrymple Island is believed by Prof. Cham-berlin never to have been covered with ice, though its altitude is now less than that of its neighbouring glaciers. In short, there is a driftless area on the northern border of Baffin Bay presenting, though on a smaller scale, the same characteristic contrasts to the region around it which appear in the driftless area of southwestern Wis-consin. From this Prof. Chamberlin infers " that the former elevation of Greenland was not coincident with conditions favouring glaciation." [*]

Several considerations, however, would seem to diminish confidence in this conclusion. In the first place, the glaciation of the Carey Islands shows that there was an immense actual enlargement of neighbouring glaciers without corresponding effects being produced upon Dalrymple Island and the neighbouring coast. This in itself calls for some local explanation, whether we suppose the glaciation to have occurred during a period of elevation or while the land was at its present comparative level. In the second place, the effect of deep depressions, such as occur in Baffin Bay and in the gulfs and fiords leading into it, upon the direction of the ice streams and upon their consequent erosive power, is not easy to estimate. There can be little doubt that, during the maximum extent of the ice in the Glacial period in the United States, the icefields surrounding the driftless area of Wisconsin were far higher than was the enclosed unglaciated region. As in the progress of great currents of water there are eddies and lines of slack motion, so there seem to have been corresponding areas of stagnant ice in the midst of continental glaciers, where the supply of foreign ice is so diminished that the melting power of the sun is adequate to the maintenance of perpetual "glacial gardens."

At the same time it must be confessed that elevation alone is not sufficient to account for extensive glaciation. The supply of moisture to the currents of air is equally necessary, for the clouds can only precipitate what they have elsewhere absorbed. It is therefore necessary to keep constantly in mind the possible influence of

---

[*] Bulletin of the Geological Society of America, vol. vi, p. 220.

changes of level in places distant from the region upon which the precipitation is taking place. The effect of an elevation in Greenland might be neutralized by an elevation in the vicinity of Behring Strait, limiting the flow of warm water into the Arctic Sea.

But the general coincidence of elevation just preceding the Glacial period is so extensive as to connect this irresistibly with the causes of the period. This evidence appears not only in the fiords and deep submerged river channels on the borders of the continent, but is equally impressive in the interior, where my own investigations have been mainly prosecuted. The inner gorges of the Ohio River and its tributaries, for example, are from three to six hundred feet in depth, and still retain their nearly precipitous sides, showing that they have been worn with great comparative rapidity during a recent geological period of continental elevation.

In the vicinity of Warren, Pa., indubitable evidence was adduced by the joint observations of Mr. Frank Leverett and myself that this erosion of the Allegheny Valley had reached certainly to a close approximation of its present extent before the first maximum period of glaciation, for we found buried channels of southern tributaries to the Allegheny filled with the very oldest glacial *débris* to a depth which carried them down very closely to the level of the present Allegheny River.*

Prof. E. H. Williams has likewise adduced a large amount of indubitable evidence that the Lehigh River in eastern Pennsylvania, and by consequence the Delaware River, into which it empties, had worn channels down nearly to the level of their present rock bottom

* See American Journal of Science, March, 1894, pp. 166–168.

before the maximum extension of ice during the first stage of the Glacial period. *

Such extensive coincidences of elevation immediately preceding the earlier stages of the Glacial period, and of the subsequent enlargement of glacial phenomena, can not easily be set aside either by *a priori* theories concerning the improbability of such extensive coincident epeirogenic changes of level as the theory supposes, or by laying stress upon numerous local phenomena of obscure import.

These oscillations of level, connected with the alternate expansion and recession of the Greenland ice, go further than some are willing to admit to sustain the theory that the Glacial period in its vicissitudes was caused by them. Indeed, the elements of the Greenland problem are essentially the same as those of the whole Northern hemisphere during the Glacial age. Prof. Shaler has warned us that New England at the present time barely escapes glacial conditions. The rudiments of a glacier still remain in Tuckermann's Ravine upon Mount Washington. A slight lowering of temperature or a slight increase of snowfall would again start the glaciers of the White Mountains out upon their career, and, when once started, it is difficult to tell where they would stop; for glaciers intensify the conditions to which they owe their origin, and would seem to have almost unlimited power when once the forces producing them have come fully into play. Equally close is the approach to glacial conditions in Norway and Alaska. Indeed, as has been shown, the oscillations of the Alaskan glaciers have been very extensive during the past century, while the delicacy of balance of cli-

---

* See American Journal of Science, January, 1894, p. 35.

matic conditions in the Alpine region is such that much
alarm was felt in Switzerland over the supposed prospect (which, however, was without foundation) that
considerable areas of the Desert of Sahara were to be
inundated by an artificial canal.   Even a slight addition, through the enlargement of the evaporating
area over which they come, to the moisture borne by
the winds which bathe the heights of the Alpine
peaks, would so increase the size of the Swiss glaciers
as to make them desolating and destructive in the
extreme.

That there have been extensive oscillations in the
relative elevation of the lands of the northern hemisphere in recent geological times is clearly proved by
the abundant facts already narrated.   But so intimately
connected are the geologic forces of the earth, that we
have to look far for the deeper causes producing the
definite results which we are called upon to study.   It
seems probable, as shown in the preceding chapters,
that the period of land elevation which preceded the
Glacial epoch was the culmination of long-continued
and slowly accumulating geologic forces.   The slow
contraction of the earth through its loss of heat, the
extensive transfers of sediment which had been going on
for ages from the elevated land areas to the sea margins,
all conspired to produce that marvellous readjustment
of the earth's surface which took place in connection
with the mountain-building era of Late Tertiary times.
It is at present impossible for us to say why the Sierra
Nevada, the Rocky Mountains, the Pyrenees, the Alps, and
the Himalayas, should have marked the lines of excessive crumpling and elevation of the earth's crust during
this period, or why the northern part of North America
and of Europe and of the intervening Atlantic basin

FIG. 62.—Looking eastward from Cape Charles, Labrador, showing the subdued character of the sky line.

should have risen so much higher than the southern portions did.

But when once these areas had arisen sufficiently high to retain throughout the season and from year to year a portion of the winter's snow, it is easy to see that a force of tremendous significance was set in operation. In due time several million cubic miles of water had been abstracted from the ocean (lowering its level fully a hundred and fifty feet) and locked up in the glacial mantle enveloping the northern part of Europe and America, a portion of which still remains in the interior of Greenland. If there is any plasticity in the earth's crust (and the geological record proves beyond question that there is), the transfer of so much weight from the oceanic beds to a limited portion of the continental area would seem sufficient to produce marked results in depressing the regions over which the ice had accumulated. After the depression connected with this cause had proceeded beyond a certain point, the lowering of the altitude would seem naturally to be an important contributing cause to bring about the change of climatic conditions which removed the ice sheet; while at the same time the removal of the burden of ice from the glaciated area and the return of the water to the oceans would partially restore the original preglacial conditions, and set in motion again the forces tending to produce a glacial period.

Thus it would seem that, in the disturbances which brought on the Glacial period, there is revealed a set of forces calculated to produce such a series of oscillations as would account partially for the complicated character of the glacial deposits, with the frequent temporary retreats and advances of the ice over extensive territories. This is the more plausible, since Greenland, too, is not without evidence of strange vicissitudes in glacial history. But, while the evidence is clear that in southern Greenland the ice formerly extended everywhere down to the ocean, it seems likely that within historical times a considerably larger area was free from ice than at the present time. For, as already remarked, it scarcely seems

Fig. 63.—Near view of the Devil's Dining Table, Labrador. (See page 36.)

possible that Norse colonies should flourish in southern Greenland under present conditions as they did nearly a thousand years ago; but at the same time, with this apparent advance of ice in southern Greenland, there is an apparent retreat in some of the northern portions. The glaciers about Inglefield Gulf, in the vicinity of Peary's headquarters, seem to be stationary, if not actually retreating.

But if we advance further, and permit ourselves to speculate upon the causes which have produced the change of level, of which there is evidence in the glaciated region, we are compelled to be satisfied with very imperfect and partial solutions of the problem. Great as is the absolute force brought to light in the alternate transfer of water from the ocean to the continental glaciers, and from the glaciers back to the ocean, its effects are disguised and modified by other forces with which they combine to such an extent that we are soon lost in endless complications. We know so little about the viscosity of the interior of the earth and about the rigidity of its crust, and about the contending forces of elevation and depression connected with the readjustments of late Tertiary times, that we must leave the subject to entice and baffle physicists and mathematicians for a long time to come. The wonders of Nature are not likely soon to be exhausted. Its secrets are not to be taken by sudden assault.

Even a slight contact with aboriginal life in Greenland makes it easy to believe in the evidence of glacial man in Europe and America. Indeed, here we have him still in Greenland, though probably of a very distinct race from the tribes which at one time hunted the walrus and the reindeer along the front of the ice sheet in the Delaware Valley, or associated with the musk ox, the cave bear, and the woolly rhinoceros on the plains of northern France. If man can maintain a subsistence along the border of the existing ice sheet of Greenland, where he is dependent for wood upon the scanty supply brought by ocean currents from Siberia, and where there is absolutely no vegetable food except a few berries and stalks of angelica, and where there are no caves and rock shelters in which he can protect himself from the inclemency of the season, it is easy to believe that he lived in comparative comfort in the well-wooded region which bordered the ice sheet on the plains of Germany and in the valley of the Mississippi, and which opened out toward more congenial climes to the south.

Fig. 64.—Contented Eskimos.

The question, however, arises, Why do men cling to such conditions of life when more attractive ones are opened to them? In Greenland, indeed, the natives are shut up to these conditions, but it would seem that the Eskimos in British America and Alaska would be attracted southward. The answer probably is, that the regions to the south are already occupied by other and

26

hostile tribes, for it is a significant fact that everywhere the Eskimo and the red Indian are inimical to each other. A distinguished botanist has said, in illustration of the truth of the doctrine of natural selection, that plants do not live where they like best, but where other plants will let them. So it seems to be, or to have been, with the Eskimos—they have evidently adjusted themselves from necessity to these northern conditions; but, once adjusted, they are now really in love with them, and Greenland life is so attractive that the race would not be at home anywhere else.

To a considerable extent the adjustment is physical. The stomachs of the Eskimos have become fitted for the digestion and assimilation of the food which the region provides, and their systems inured to the kind of labour necessary to provide the requisite food and raiment. The missionaries and Danish officials have found it undesirable, and indeed impossible, greatly to change the habits of native life in Greenland and Labrador; for, as already said, the value of native products is greater in use than it can be in exchange. The blubber and skin of the seal meet a more important want to the Greenlander himself than they possibly can to civilized nations; while the food and clothing of temperate climates are of little service in the rigours of a Greenland winter or amid the exposures of the hunting season in Greenland waters; and a European house, fitted to protect against the cold and storms of an arctic region, would be so much more expensive than the house of the natives as to be entirely beyond their means.

Life in Greenland, however, is by no means so dreary as it would seem to an outsider, and one can well understand how people physically adjusted to the conditions maintain such cheerfulness and gaiety in their life.

Indeed, it is not possible for the natives of southern climes to be more light-hearted than are the Greenland Eskimos. As we have seen, they are given to singing and dancing, to story-telling and joking, and their merry laughter rings out everywhere to enliven all

FIG. 65.—Eskimos sporting in their kayaks. One is jumping over the other.

ordinary intercourse. A little closer examination of the conditions, likewise, shows that there is a secure basis for this cheerfulness and hope. Up to certain limits the supply of food and raiment for the Eskimos is abundant and readily attainable. Their low houses, built of stone and sod, and covered with a slightly arching roof, are further protected during the winter by the great fall of snow beneath which they are deeply buried. In these houses so protected the flame of the oil lamp furnishes all the heat desirable; so that for comfort the inhabitants are compelled to divest themselves of most of their clothing when they enter the room, and thus are the better protected by it when they return again to their duties out of doors. Even the venerable Dr. Rink testifies that the welcome shelter of such a house is not without its attractions when the storm is raging outside.

The supply of food and raiment is brought to the very door of the Eskimos in Greenland, or perhaps it is more proper to say that they have settled at those places where these natural supplies are most constant and abundant; thus illustrating the same principle which has led to the common observation that usually Nature has so ordered it that there should be a bountiful run of salmon in the vicinity of a monastery, and that a large river should flow by a great city. So the supply of natural advantages in Greenland, though limited, is remarkably uniform and reliable. Indeed, it is so constant as to interfere with the development of forethought, prudence, and thrift among the people, and to stand in the way of their progress and advancement, and to favour the continuance of that communistic life which characterizes their state of society.

Instead of having to go out to distant seas, as the hardy fishermen of Newfoundland and Norway do, for their fish and seal, the fish and seal come to the Greenlanders. And, indeed, in a country where there is no ship timber, it is impossible to think of venturing far from the coast. The Greenlander, therefore, must abide his time and wait till the seasons for the schools of fish

arrive, and for the annual migration of the various species of seal, and of the immense flocks of birds, whose
eggs and flesh give pleasing variety of food and whose
feathers afford unrivalled protection against the cold.
The very regularity of this supply removes, as we have
said, the motives which in a state of civilization lead to
provision for the future, and which stimulate and diversify the activities of men.

But, at the same time, it is clear that the population
of Greenland is narrowly limited in growth by the average amount of these annual supplies. It is not possible
for the population of Greenland largely to increase.
Much increase in the destruction of the animal life
upon which the people depend for their existence would
" kill the goose that lays the golden egg," and lead to
a speedy diminution of resources. Indeed, as we have
already remarked, owing to the facilities furnished by
firearms for killing reindeer a generation ago, and the
inducement held out by even the low exchangeable
value of the horns and hides and tongues of those animals, and a failure upon the part of the Government to
enforce proper restrictions, the whole species of reindeer
has been brought to the verge of extinction in Greenland; so that not only great hardships are inflicted
upon the present generation, but total destruction of
the Greenland Eskimo is threatened.

For, as remarked in the previous chapter, while the
Greenland igloo is not seriously objectionable from a
sanitary point of view during the winter months, when
everything is frozen so as to prevent decay, nothing can
exceed the filthiness of the surroundings when the snows
melt off in the spring. The evils of this unsanitary condition were formerly avoided by the possibility of each
family's possessing a reindeer-skin tent, in which they

took up their abode for the summer, while they left the elements to purify their winter abodes. But, now that the supply of reindeer skins is so limited, the inhabitants are compelled to occupy their unsanitary abodes during the summer season, greatly to the detriment both of their own health and of that of their posterity. In this case certainly the possession of civilized instruments of destruction has proved a doubtful blessing to the natives of Greenland. How far the wise foresight of the estimable Danish officials and missionaries may correct the present evil tendencies is one of the most interesting questions of the future. Happily, the problem in Greenland is much simpler than that which is presented in the preservation of the aborigines of our own country; for in Greenland there is no inducement for civilized races to encroach upon their possessions and subject the natives to the irrepressible conflicts which are so continuously destroying the Indian tribes in the United States. It being impossible to transport European civilization to the conditions of Greenland, it is not necessary to waste our sympathy upon the people who are already adjusted to these conditions and are happy in them. If any people live in Greenland, they must live substantially as the Eskimos do. We may be content if we successfully impart to them a share in that higher intellectual and spiritual life which ennobles and sweetens all conditions alike.

A similar line of remarks is suggested by the conditions of life throughout the entire North Atlantic region. Scanty as is the population of Labrador and Newfoundland, it has about reached its maximum limit, unless other than the present sources of subsistence are discovered and utilized. The seals which annually float past the coast of Labrador are not unlimited in number, and are not wholly insensible to the successive inroads made upon them by modern methods of destruction. There is imminent danger that the numbers

Fig. 66.—Towing the Rigel out of the Punch Bowl during a calm.

shall be so reduced by the present greed of hunters that there will be little reward awaiting the enterprising crews who set out in the future from Newfoundland for their capture. Indeed, the recent stress of hard times in Newfoundland is in part a result of the failing supply of seal to be found upon the ice of the Labrador current.

The impressive regularity with which the schools of different kinds of fish visit stated places from year to year throughout the North Atlantic furnishes a solid basis for business calculation up to a certain point.

But evidently there is a limit to the supply of available fish, even in the ocean, and the increase in the catch is by no means proportionate to the increase in the numbers and efforts of the fishermen. It would not by any means be so easy to double the quantity of fish caught as it is to double the means for their capture.

One can not, however, but be deeply impressed with the success of the people of the North Atlantic in adjusting themselves to the conditions surrounding them. The few thousand people who remain in Labrador seem to fill the space which Nature has provided, and are able to maintain an existence that is by no means without its attractions. The winters are indeed long, but not altogether dreary, since with their dog sledges the inhabitants can travel over the frozen bays more readily than they can cross them by boat in summer; while the moderate supply of timber, which is protected by its poor quality from total destruction by lumbermen, supplies them with shelter and fuel and a limited amount of occupation. The difficulty of securing educational and religious facilities is partly overcome by gathering close together in winter time, and by the use of floating bethels in the summer; while the great influx of foreign ships and outside fishermen during the summer effectually enlarges the horizon of the native's mental vision.

Contact both with the permanent European residents of Labrador and with the captains and crews of the vessels which venture into these northern regions to supply the world's demand for fish, can but greatly increase one's appreciation of the marvellous capacity of human nature for adapting itself to seemingly adverse conditions, and for wresting from them by conflict the noblest qualities of character. The Gloucester fishermen whom one meets on the coasts of Iceland and

Greenland, and the Newfoundlanders whom he encounters upon the coast of Labrador, as well as those who are permanent residents there, are easily the peers of any class of people in the world. Perhaps, however, this is a result of that stern law of natural selection which here operates with remorseless certainty, permitting only the most healthy, the most enterprising, and the most intelligent to survive.

تتمه

INDEX.

INDEX.

Carpenter, P. P., 317, 318.
Cartensen, A. R., 96, 98.
Causes of the Ice Age, 211, 215, 305, 334–347.
Cetaceans, 222, 233–236.
Chamberlin, Prof. T. C., cited, 200, 269, 306, 308, 336, 341; observations of Greenland glaciers, 287–293, 374; on stages of the Ice Age, 349, 351.
Champlain, Lake, 326, 327, 345, 351, 359.
Champlain epoch, 233, 235, 307, 309, 325, 345, 351, 360; subsidence, 322–330, 345, 355, 358.
Chandler, Dr. S. C., 343.
Chapels in Labrador, 25, 27, 28, 32.
Charles Harbour, Cape, vi, 13, 23, 25, 27, 31.
Chateau Bay, 35, 45.
Chimo, Fort, 44.
Chinese, 134 et seq.
Chinook winds of the United States, 123.
Christianshaab, district of, 109.
Cinquefoil, 191, 194, 197, 198.
Circumpolar fauna, 215; flora, 201, 203, 206.
Claushavn, 109.
Climatic changes in Greenland, 307, 311, 331.
Cloudberry, 30, 192.
Club moss, 197.
Coal in Greenland, 111, 113, 207; Grinnell Land, 210.
Codfish, 27, 31, 32, 87, 106, 109, 240, 242.
Compositæ, 196.
Coniferæ in Greenland, 191; fossil, 208, 210, 211.
Coniferous driftwood, 323, 324.
Cook, Dr. F. A., v, x, 50, 136, 150.
Cotton grass, 193.
Cowberry, 192.

Cress family, 195, 196.
Cretaceous strata in Greenland, 207, 209, 213, 319.
Crevasses, 92, 253, 260, 264, 266, 271, 274, 278, 299.
Crimson Cliffs, 200.
Croll, Dr. James, 335–343.
Crowberry, 109, 189, 199.
Crowfoot, 191, 196.
Cruciferæ, 195, 196.
Crustacea, 244.
Cryolite mines, 15, 70, 101, 185.
Cryptogamous plants, 194, 197, 200.
Curlew berry, 36, 189
Cushing, Prof. H. P., 301.
Cycads, fossil, 209.

Dalager, Lars, on inland ice, 103, 252, 265.
Dalager's nunataks, 252, 264, 373.
Dale, Dr. William H., 188, 333.
Dalrymple Island, 292, 374, 375.
Dana, Prof. J. D., 264, 266, 304, 306, 312, 315, 337, 344, 345, 351.
Danes in Greenland, explorations, 250, 251, 263, 266, 285, 328; fuel, 193.
Dannebrog's Island, 118.
Darwin, Charles, 48, 317, 346.
Davidson, Prof. George, 318.
Davis, Dr. G. G., 280.
Davis Strait, 28, 46, 54; depth, 313; geology of sides of, 29; limit of migration of plants, 202, 205, 207, 369.
Dawson, Dr. George M., 304, 319, 352.
Dawson, Sir William, 210, 235, 315, 321, 325.
Decomposition of gneiss, 292.
De Geer, Baron Gerard, 329, 330, 336, 349.
Delaware Bay, submarine fiord, 315. River, 376.

27

THE END.

*C*LIMBING IN THE HIMALAYAS. By WILLIAM
MARTIN CONWAY, M. A., F. R. G. S., Vice-President of the Alpine Club; formerly Professor of Art in University College,
Liverpool. With 300 Illustrations by A. D. McCORMICK, and
a Map. 8vo. Cloth, $10.00.

This work contains a minute record of one of the most important and
thrilling geographical enterprises of the century—an expedition made in
1892, under the auspices of the Royal Geographical Society, the Royal
Society, the British Association, and the Government of India. It included
an exploration of the glaciers at the head of the Bagrot Valley and the great
peaks in the neighborhood of Rakipushi (25,500 feet); an expedition to
Hispar, at the foot of the longest glacier in the world outside the polar
regions; the first definitely recorded passage of the Hispar Pass, the longest
known pass in the world; and the ascent of Pioneer Peak (about 23,000
feet), the highest ascent yet authentically made. No better man could have
been chosen for this important expedition than Mr. Conway, who has spent
over twenty years in mountaineering work in the Alps. Already the author
of nine published books, he has recorded his discoveries in this volume in the
clear, incisive, and thrilling language of an expert.

"It would be hard to say too much in praise of this superb work. As a record of
mountaineering it is almost, if not quite, unique. Among records of Himalayan exploration it certainly stands alone. . . . The farther Himalayas . . . have never been
so faithfully—in other words, so poetically—presented as in the masterly delicate
sketches with which Mr. McCormick has adorned this book."—*London Daily News.*

"This stately volume is a worthy record of a splendid journey. . . . The book is
not merely the narrative of the best organized and most successful mountaineering expedition as yet made; it is a most valuable and minute account, based on first-hand
evidence, of a most fascinating region of the heaven-soaring Himalayas."—*Pall Mall
Gazette.*

"Mr. Conway's volume is a splendid record of a daring and adventurous scientific
expedition. . . . What Mr. Whymper did for the Northern Andes, Mr. Conway has
done for the Karakorum Himalayas."—*London Times.*

"It would be difficult to say which of the many classes of readers who will welcome
the work will find most enjoyment in its fascinating pages. Mr. Conway's pen and Mr.
McCormick's pencil have made their countrymen partners in their pleasure."—*London
Standard.*

". . . In addition to this, Mr. Conway is a man of letters, a student (and a teacher,
too) of art, a scholar in several languages; one, too, who knows the Latin names of
plants, and the use of theodolite and plane table. From him, therefore, if from any
one, the world had a right to expect a book that should combine accurate observation
and intelligible reporting with an original and acute record of impressions; nor will
the world have any reason to be disappointed."—*London Athenæum.*

"With its three hundred illustrations we have seldom seen a volume which speaks
to the eye and understanding so pleasantly and expressively on every page. . . . We
have an exhaustive panorama of the Himalayan scenery, of the manner in which the
rough marching was conducted, of ascents achieved under the most dangerous conditions, and of the troubles and humors of the shifting camps where the coolies rested
from their labors."—*London Saturday Review.*

"Perhaps no book of recent date gives a simpler or at the same time more effective
picture of the truly wonderful mountain regions lying behind the northern barrier of
India than Mr. Conway's striking volume."—*London Telegraph.*

THE ICE AGE IN NORTH AMERICA, and its Bearings upon the Antiquity of Man. By G. FREDERICK WRIGHT, D. D., LL. D., F. G. S. A., Professor in Oberlin Theological Seminary; Assistant on the United States Geological Survey. With an appendix on "The Probable Cause of Glaciation," by WARREN UPHAM, F. G. S. A., Assistant on the Geological Surveys of New Hampshire, Minnesota, and the United States. New and enlarged edition. With 150 Maps and Illustrations. 8vo, 625 pages, and Index. Cloth, $5.00.

"Not a novel in all the list of this year's publications has in it any pages of more thrilling interest than can be found in this book by Professor Wright. There is nothing pedantic in the narrative, and the most serious themes and startling discoveries are treated with such charming naturalness and simplicity that boys and girls, as well as their seniors, will be attracted to the story, and find it difficult to lay it aside."—*New York Journal of Commerce.*

"One of the most absorbing and interesting of all the recent issues in the department of popular science."—*Chicago Herald.*

"Though his subject is a very deep one, his style is so very unaffected and perspicuous that even the unscientific reader can peruse it with intelligence and profit. In reading such a book we are led almost to wonder that so much that is scientific can be put in language so comparatively simple."—*New York Observer.*

"The author has seen with his own eyes the most important phenomena of the Ice age on this continent from Maine to Alaska. In the work itself, elementary description is combined with a broad, scientific, and philosophic method, without abandoning for a moment the purely scientific character. Professor Wright has contrived to give the whole a philosophical direction which lends interest and inspiration to it, and which in the chapters on Man and the Glacial Period rises to something like dramatic intensity."—*The Independent.*

". . . To the great advance that has been made in late years in the accuracy and cheapness of processes of photographic reproduction is due a further signal advantage that Dr. Wright's work possesses over his predecessors'. He has thus been able to illustrate most of the natural phenomena to which he refers by views taken in the field, many of which have been generously loaned by the United States Geological Survey, in some cases from unpublished material; and he has admirably supplemented them by numerous maps and diagrams."—*The Nation.*

MAN AND THE GLACIAL PERIOD. By G. FREDERICK WRIGHT, D. D., LL. D., author of "The Ice Age in North America," "Logic of Christian Evidences," etc. International Scientific Series. With numerous Illustrations. 12mo. Cloth, $1.75.

"It may be described in a word as the best summary of scientific conclusions concerning the question of man's antiquity as affected by his known relations to geological time."—*Philadelphia Press.*

"The earlier chapters describing glacial action, and the traces of it in North America—especially the defining of its limits, such as the terminal moraine of the great movement itself—are of great interest and value. The maps and diagrams are of much assistance in enabling the reader to grasp the vast extent of the movement."—*London Spectator.*

New York: D. APPLETON & CO., 72 Fifth Avenue.

CAMP-FIRES OF A NATURALIST. From the Field Notes of LEWIS LINDSAY DYCHE, A. M., M. S., Professor of Zoölogy and Curator of Birds and Mammals in the Kansas State University. The Story of Fourteen Expeditions after North American Mammals. By CLARENCE E. EDWORDS. With numerous Illustrations. 12mo. Cloth, $1.50.

"It is not always that a professor of zoölogy is so enthusiastic a sportsman as Prof. Dyche. His hunting exploits are as varied as those of Gordon Cumming, for example, in South Africa. His grizzly bear is as dangerous as the lion, and his mountain sheep and goats more difficult to stalk and shoot than any creatures of the torrid zone. Evidently he came by his tastes as a hunter from lifelong experience."—*New York Tribune.*

"The book has no dull pages, and is often excitingly interesting, and fully instructive as to the habits, haunts, and nature of wild beasts."—*Chicago Inter-Ocean.*

"There is abundance of interesting incident in addition to the scientific element, and the illustrations are numerous and highly graphic as to the big game met by the hunters, and the hardships cheerfully undertaken."—*Brooklyn Eagle.*

"The narrative is simple and manly and full of the freedom of forests. . . . This record of his work ought to awaken the interest of the generation growing up, if only by the contrast of his active experience of the resources of Nature and of savage life with the background of culture and the environment of educational advantages that are being rapidly formed for the students of the United States. Prof. Dyche seems, from this account of him, to have thought no personal hardship or exertion wasted in his attempt to collect facts, that the naturalist of the future may be provided with complete and verified ideas as to species which will soon be extinct. This is good work—work that we need and that posterity will recognize with gratitude. The illustrations of the book are interesting, and the type is clear."—*New York Times.*

"The adventures are simply told, but some of them are thrilling of necessity, however modestly the narrator does his work. Prof. Dyche has had about as many experiences in the way of hunting for science as fall to the lot of the most fortunate, and this recountal of them is most interesting. The camps from which he worked ranged from the Lake of the Woods to Arizona, and northwest to British Columbia, and in every region he was successful in securing rare specimens for his museum."—*Chicago Times.*

"The literary construction is refreshing. The reader is carried into the midst of the very scenes of which the author tells, not by elaborateness of description but by the directness and vividness of every sentence. He is given no opportunity to abandon the companions with which the book has provided him, for incident is made to follow incident with no intervening literary padding. In fact, the book is all action."—*Kansas City Journal.*

"As an outdoor book of camping and hunting this book possesses a timely interest, but it also has the merit of scientific exactness in the descriptions of the habits, peculiarities, and haunts of wild animals."—*Philadelphia Press.*

"But what is most important of all in a narrative of this kind—for it seems to us that 'Camp-Fires of a Naturalist' was written first of all for entertainment—these notes neither have been 'dressed up' and their accuracy thereby impaired, nor yet retailed in a dry and statistical manner. The book, in a word, is a plain narrative of adventures among the larger American animals."—*Philadelphia Bulletin.*

"We recommend it most heartily to old and young alike, and suggest it as a beautiful souvenir volume for those who have seen the wonderful display of mounted animals at the World's Fair."—*Topeka Capital.*

New York: D. APPLETON & CO., 72 Fifth Avenue.

THE STORY OF THE WEST SERIES.

Edited by Ripley Hitchcock.

"There is a vast extent of territory lying between the Missouri River and the Pacific coast which has barely been skimmed over so far. That the conditions of life therein are undergoing changes little short of marvelous will be understood when one recalls the fact that the first white male child born in Kansas is still living there; and Kansas is by no means one of the newer States. Revolutionary indeed has been the upturning of the old condition of affairs, and little remains thereof, and less will remain as each year goes by, until presently there will be only tradition of the Sioux and Comanches, the cowboy life, the wild horse, and the antelope. Histories, many of them, have been written about the Western country alluded to, but most if not practically all by outsiders who knew not personally that life of kaleidoscopic allurement. But ere it shall have vanished forever we are likely to have truthful, complete, and charming portrayals of it produced by men who actually know the life and have the power to describe it."— *Henry Edward Rood, in The Mail and Express.*

NOW READY.

THE STORY OF THE INDIAN. By George Bird Grinnell, author of " Pawnee Hero Stories," " Blackfoot Lodge Tales," etc. 12mo. Cloth. Illustrated. $1.50.

" A valuable study of Indian life and character. . . . An attractive book, . . . in large part one in which Indians themselves might have written."—*New York Tribune.*

"Among the various books respecting the aborigines of America, Mr. Grinnell's easily takes a leading position. He takes the reader directly to the camp-fire and the council, and shows us the American Indian as he really is. . . . A book which will convey much interesting knowledge respecting a race which is now fast passing away."—*Boston Commercial Bulletin.*

" It must not be supposed that the volume is one only for scholars and libraries of reference. It is far more than that. While it is a true story, yet it is a story none the less abounding in picturesque description and charming anecdote. We regard it as a valuable contribution to American literature."—*N. Y. Mail and Express.*

" A most attractive book, which presents an admirable graphic picture of the actual Indian, whose home life, religious observances, amusements, together with the various phases of his devotion to war and the chase, and finally the effects of encroaching civilization, are delineated with a certainty and an absence of sentimentalism or hostile prejudice that impart a peculiar distinction to this eloquent story of a passing life."—*Buffalo Commercial.*

" No man is better qualified than Mr. Grinnell to introduce this series with the story of the original owner of the West, the North American Indian. Long acquaintance and association with the Indians, and membership in a tribe, combined with a high degree of literary ability and thorough education, has fitted the author to understand the red man and to present him fairly to others."—*New York Observer.*

IN PREPARATION.

The Story of the Mine. By Charles Howard Shinn.
The Story of the Trapper. By Gilbert Parker.
The Story of the Explorer.
The Story of the Cowboy.
The Story of the Soldier.
The Story of the Railroad.

www.ingramcontent.com/pod-product-compliance
Lightning Source LLC
Chambersburg PA
CBHW021345210326
41599CB00011B/760